# WHAT IS EXTINCTION?

# WHAT IS EXTINCTION?

A Natural and Cultural History
of Last Animals

JOSHUA SCHUSTER

Fordham University Press
NEW YORK   2023

Fordham University Press has no responsibility for the persistence or accuracy of URLs for external or third-party Internet websites referred to in this publication and does not guarantee that any content on such websites is, or will remain, accurate or appropriate.

Fordham University Press also publishes its books in a variety of electronic formats. Some content that appears in print may not be available in electronic books.

Visit us online at www.fordhampress.com.

Library of Congress Cataloging-in-Publication Data available online at https://catalog.loc.gov.

Printed in the United States of America

25  24  23    5  4  3  2  1

First edition

# Contents

# WHAT IS EXTINCTION?

# Introduction

Take a moment to consider the current population numbers of a few iconic mammals. Today, there are approximately 26,000 polar bears, 20,000 lions, 7,000 cheetahs, and 10,000 blue whales.[1] There are 3,000 wild tigers, but more than 10,000 tigers in captivity. There are 30,000 rhinos across five different subspecies, compared to numbers in the millions a few centuries ago. For the subspecies of northern white rhinoceros, there remain only two females and no males. Among the great apes, there are 300,000 western gorillas and 100,000 orangutans, and these populations are classified as critically endangered because they have diminished by over 50 percent in the past few decades and their numbers continue to decline.

A 2018 global study of wildlife led by World Wildlife Fund found that animal populations across the planet have diminished by 60 percent between 1970 and 2014.[2] The poet Camille Dungy puts it succinctly: "No lions, no tigers, no bears."[3] Similar massive, sweeping devastations of insects (over 40 percent of insects are threatened with extinction),[4] birds,[5] fish,[6] and plants[7] has also occurred in recent decades.

When I first learned of these numbers a few years ago (the populations have not changed much since then), I felt staggered and deserted. I had not known that death had undone so many so recently. I began asking people around me—friends, family members, strangers—to guess how many wild lions and tigers currently existed. Such melancholic knowledge, I found, turns one into a kind of Coleridgean ancient mariner. One cannot help but stop people and compulsively tell the story of last animals. In Coleridge's *The Rime of the Ancient Mariner*, the grizzled mariner stops the next of kin of the bridegroom from entering the church and participating in the wedding ceremony. The mariner tells his story of abusing the omen of the albatross, killing the bird, and casting his cursed ship into death-ridden infamy.

The wedding guest is waylaid on the threshold of an event celebrating the ongoing rituals of human matrimony, kinship, and reproductive life. By contrast, the mariner is endlessly, hauntingly driven to repeat his story of the death of his crew initiated by his disrespect of the ocean and its animal inhabitants. The mariner's terse final admonition to the wedding guest—"He prayeth well, who loveth well / Both man and bird and beast"[8]— is unsettled by the cursed mariner's realization that life and death for humans and animals are no longer separate or distinct realms. To pray and love well must be experienced as the imbrication of humans and animals who share the same blessings and curses of love, forlornness, terror, and finitude. It turns out Coleridge's poem also has become an augury of the story of extinction for the albatross. According to the International Union for the Conservation of Nature (IUCN) Red List assessments, all the albatross populations are listed as "concerned," with several subspecies classified as critically endangered due to human-induced causes, including entanglement in fishing gear, over-fishing, hunting, and habitat destruction.[9]

This is what it is like to talk of extinction today—it makes one into a spectral and captivated mariner-like figure, stopping one of three, telling the waylaid the enthralling and ghastly story of what it means for a species to disappear in the midst of what appears to be a time of plenty. Like the Coleridge poem, the teller needs to use art and science to speak about the limits of life that also anguish the limits of the mind. And also like the mariner, the compulsive, relentless telling of the story of extinction will lead to questioning everything about the means of storytelling itself. The scientific confirmation of the fact of species extinction, first made at the end of the eighteenth century (the same moment as Coleridge's poem), revealed a new kind of loss, not just the death of an animal but the end of an entire way of life.

The knowledge of extinction has permanently changed definitions of both life and death. It has given rise to efforts to conceptualize and register something more radical than death: the apprehension of nonexistence and futurelessness. The end of a species is an emptiness in being itself, a diminishing of reality, and a permanent hole in what Charles Darwin called the "polity of nature."[10] Today, estimates indicate over 8 million species currently inhabit the planet.[11] To tell the story of this wonderful profusion of life would require a dazzling and astonishing narrative scope.

Yet it is just as astonishing to realize that telling this story also includes the revelation that, eventually, every one of these species will, at some point, disappear.

How to begin to tell such stories? There is no clock that tells us when a given species will become extinct; yet, statistically, there are measurements of the average rate of extinction and also measurements of extreme accelerations in extinction, called mass extinction events. Paleontological research indicates that approximately 99.9 percent of all species in the history of our planet have gone extinct.[12] The biologist Ernst Mayr estimates that well over 1 billion species have disappeared during the history of the Earth.[13] These numbers show the devastating yet consistent, even functional, role extinction plays in the story of speciation. Research by paleontologists Jack Sepkoski and David Raup in the early 1980s using quantitative methods and computer models found that several more intense episodes of species loss, as well as five mass extinction events, have occurred in the history of life on Earth. This led David Sepkoski to suggest: "In fact, from an ecological perspective, mass extinctions may even be a *requirement* for developing complex systems like the earth's."[14]

The contributions of the "paleobiological revolution"[15] in the study of extinction in recent decades now forms the basis for new reckonings and historical consciousness on the magnitude of the ends of species. Viewed from this vantage, extinction is a fact of biological regularity apparent in the deep-time scale of species history, yet knowledge of the increasing rate of the depletion of life is also a marker of modernity and of the present. The extinction of species is one way we have come to understand both vast stretches of passing time and the precariousness of life today. The death of species can follow statistical norms and patterns, but, also, there are instances of huge fluctuations of extinction rates that cast norms of life aside. In the language of nineteenth-century biology, extinction is both uniformitarian and catastrophist. Extinction can be regulative and alarmist, functional and apocalyptic, measurable and disastrous, evolutionary and entropic, universally permanent yet, perhaps, locally reversible through new biotechnologies.

There is ample evidence that we are today living in extraordinary times of the loss of species, what has been called the "sixth mass extinction" event in Earth's 3.5 billion-year history of life.[16] While it is difficult to assert a single number that would index the rate of all extinctions today—since the demise of a species depends on several factors, including its population size, habitat range, mobility, magnitude of harms, and geographical isolation (species native to small islands or isolated locales are much more prone to extinctions than species spread across larger areas)—scientists have accumulated evidence that extinction rates are rising by a factor of 10 to 100 in many locations across the globe.[17]

Extinction rates are rising largely due to human causes, including hunting and fishing; habitat destruction; deforestation; global heating; ocean acidification; pollution; pesticides; the harvesting of rare animals as pets or for food and medicines; and the introduction of new predators to new habitats. The sweeping defaunation of wildlife populations across the globe is part of this extinction event. A stunning 2019 report from the Intergovernmental Science-Policy Platform on Biodiversity and Ecosystem Services (IPBES), compiled by 145 scientists across the globe, found that approximately 25 percent of all species are threatened.[18] The report concluded that it is possible that 1 million species will go extinct in the near future, with many lost in the coming decades, if no action is taken to halt current trends.

Today, human activity is changing the biology as well as the meaning of extinction. We live in a time of animal lastness, as evident in the plethora of last titles: *Last Chance to See, The Last Tiger, The Last Pictures, Last of the Curlews, The Last Elephants, The Last Animals, The Last Lions, The Last Polar Bear, The Last Panda, The Last Rhinos.*[19] The extinction of any species—from the American chestnut moth to the Caribbean monk seal to the passenger pigeon—effectuates a permanent change in the scope and space of existence. These accounts of animal extinctions resonate with the proliferation of cultural works envisioning a world without humans. The biological end of many species has become palpable as an everyday concern. We are the human witnesses to a moment where humans have put existence itself into question. In placing all of life into question, we have placed, also, our own lives into question. There is no better time to raise the question of "What is extinction?" and to pose this inquiry in such a way that it does not reproduce answers tied to the current diminishment of lifeworlds. The aim of this book is to examine the scenes and rhetorics of lastness for species while resisting the colonialist and dominatory assertions of "firsting and lasting" that Indigenous historian Jean M. O'Brien (White Earth Ojibway) identifies as pervasive in settler science and culture predicated on doctrines of discovery declaring "first" claims and announcing "last" existences.[20]

To understand how the significance of extinction is being redefined today, we need to reflect on previous shifts and turning points in the knowledge of extinction over the 250 years since the first scientific descriptions of species finitude. The chapters in this book focus on several prominent events of animal and human extinction in the past two centuries that have effectuated dramatic changes in public awareness of the disappearance of species and the concepts used to cope with their permanent loss. It covers the period of the first reactions to the scientific understanding of extinction by naturalists who were precursors to Darwin, and includes analysis up to the present situation, in which a global consensus has formed of the need to confront the precipitous disappearance of species and animal populations. My objective is to provide "close readings" of extinction events. I discuss case studies here of both animal and human extinction, including historical instances of genocide that were attempts by some humans to condemn others to oblivion. These chapters proceed with a cautious regard to how the history of the knowledge of extinction at times has required the conflation of concepts of animality and humanity toward a shared sense of finitude and, other times, when this very conflation has been used to bifurcate humanity into violent hierarchies used as the basis for justifying the extinction of some human groups.

Understanding the history of the concept of extinction as applied to humans requires carefully parsing how the scientific study of extinction is entangled with the record of extinctionary acts toward both human and nonhuman lives. The empirical-based scientific knowledge that humans, as a biological species among others, eventually will go extinct seems almost an entirely different notion than genocide. Yet the history of conceptualizing extinction overlaps with the history of mass violence toward some human communities and peoples who have been subject to genocidal agendas informed by naturalist interpretations of extinction.

Nicholas Mirzoeff discusses how every single acclaimed nineteenth-century naturalist thinker about extinction also extended this science into violent judgments of the biological basis for the oppression of non-white races. "The concept of extinction itself was part of the transformation of natural history into life science (biology) in the era of the revolutions of the enslaved and abolition (1791–1863),"[21] Mirzoeff details. The science of extinction served not just to raise ethical consciousness about the finitude of life as such but also to introduce new forms of violence and dominance by marking off some species and races as more extinctable than others. Mirzoeff remarks: "No sooner had the concept of extinction been announced by Georges Cuvier in the early nineteenth century than its founder was hard at work trying to define an essential and visible 'line' or difference between Africans and Europeans" (124). The generalized existential anguish that humans as a whole might eventually face extinction coincided with a new colonial purview, one based on the ongoing destructions of some human and nonhuman animal communities whose bodies and minds had been marked as racially or socially inferior and whose entire way of life was treated as fated to be suppressed and surpassed. The circulation of concepts of extinction over the past two centuries has consistently imbricated apparently neutral empirical claims for the perishability of all species with planned violent acts to deprive the existence of some lives deemed irrecoverable and expendable.

The history of animal extinction cannot be separated from the history of human extinctionary events toward other humans, what recent researchers have emphasized as the "genocide-ecocide nexus."[22] The diminishment and destruction of animal lives over the last several centuries connects distinctly with the diminishment and destruction of Indigenous communities, in particular, that have existed in long-standing relationship with these animals. Winona LaDuke explains how the continued elimination of Indigenous peoples parallels the ongoing destruction of animal and plant species:

> There have been more species lost in the past 150 years than since the Ice Age. During the same time, Indigenous peoples have been disappearing from the face of the earth. Over 2,000 nations of Indigenous peoples have gone extinct in the western hemisphere, and one nation disappears from the Amazon rainforest every year. There is a direct relationship between the loss of cultural diversity and the loss of biodiversity.[23]

The diversity of peoples and languages mirrors the diversity of species. The disappearance of peoples historically has coincided with the disappearance of the planet's biodiverse life, resulting in what Rob Nixon describes as a "conjoined ecological and human disposability"[24] facing impoverished and marginalized communities across the globe today.

It is crucial then to note how attention to the scientific and cultural ramifications of the sixth mass animal extinction arose in a belated way, considering the much longer history of the violent colonization and domination of other humans and nonhuman species, which often pressed toward extremes of extinction. Potawatomi scholar Kyle Powys Whyte situates the recent rush to compose a sense of disastrous climate change and extinction as a situation of déjà vu for Indigenous peoples across the globe who have already experienced centuries of apocalyptic existential events. As Whyte states, "the hardships many non-Indigenous people dread most of the climate crisis are ones that Indigenous peoples have endured already due to different forms of colonialism: ecosystem collapse, species loss, economic crash, drastic relocation, and cultural disintegration."[25]

To take one example, the Indigenous peoples of Guanihaní (Haiti and Dominican Republic), the Taíno and Arawak, numbered between half a million to 1 million before Columbus arrived in 1492. Within a few decades, the Indigenous population fell to 30,000 due to disease, enslavement, and execution, and by the end of the sixteenth century, the Spanish colonists recorded that the Taíno and Arawak had been eradicated as a distinct people (although aspects of their languages, cultures, and communities still survive today).[26]

Sixteenth- and seventeenth-century colonists witnessed and often participated in processes of depopulation and displacement leading toward extinction, but intellectual histories of the concept of extinction typically begin in the late eighteenth century with the scientific discovery of fossils of extinct animals. Even in the case of the dodo, hunted into eradication by Dutch mercantilists and colonialists, its demise became apparent by the end of the seventeenth century, nearly a century before most scientific formulations on species death. The belatedness of the scientific consensus of species disappearance, finally confirmed in the early nineteenth century, provides evidence of how scientific and cultural concepts of extinction intertwine with global histories of power and regimes of possession and dispossession.

The chapters in this book provide interpretations of prominent historical moments and debates around the end of species, to show how thinking about extinction has changed in different contexts for different species, how the loss of species has been highlighted or dismissed, and how the concept of extinction has been used in situations to redefine the ends of human and nonhuman animal life. The extinction of humans often has been enacted or portrayed in dramatically different ways than the extinctions of animals over the past several centuries. However, the initial scientific and cultural reflections on extinction stemming from the late eighteenth century also recognized that extinction was a shared condition and a primordial harm that haunted all species. Extinction revealed that the essence of being a species is to eventually be deprived of existence. To think of oneself as a species among others, then, is to think of the finitude of one's species. The early natural scientists of extinction stumbled upon this paradoxical tenet of humanistic thought: to be human includes contemplating no more humans. Yet what this discovery of the finitude of species means for the past, present, and future of extinction has been contested in extreme ways, opening the gateway to unspeakable acts of violence and horror, but also to the possibility for a collective determination to share the Earth in a way that recognizes the ecological precarity of all existences.

The events of animal and human extinction that form the basis for this book also have elicited profound cultural and aesthetic responses that raise questions about what happens to artworks that attend to the limits of life. I examine a number of works of art that respond to the end of a species form while also posing the matter of what form art's own finitude could take. In the broad scheme of the magnitude of species loss and planetary environmental problems, it might seem an odd deflection to examine artworks, even those that consider the ecological urgency of the depletion of animal populations. Yet extinction is not just a biological event and is not knowable only biologically; it also is a concept and a historical event that involves multiple agencies, documentations, metaphors, and cultural forms that solicit multiple kinds of awareness. Artworks that attend to the ends of species and to the finitude of art itself also can help change the meanings of life, death, and extinction toward more conscientious and caring efforts.

From the outset, one must speak of different senses of *extinctions* circulating in the plural at different historical moments. The specificity in time and place of extinction—Extinction when? Extinction where?—matters centrally for what kinds of representations and public cultures adhere to such scenes. In examining the when and where of extinction, for each event, it matters who is disappearing and who is observing. It matters if we are speaking of mammals, of insects, of plants, or of single-cell or multicellular organisms, all of whose extinctions are figured in distinctly different conceptual registers and ecological terms. Definitions and cultures of extinction turn on which animals and which humans or human communities are figured as extinct or prone to extinction, who witnesses, who gets to speak from the third-person plural "we," and from what sense of perspective and knowledge. The kinds of language, stories, styles, points of view, and conceptual terms used to document and comprehend the end of a species form contribute to the meanings of extinction. As Ursula Heise writes, "engagements with endangered species depends on these broader structures of imagination, and individuals' paths to conservationist engagement become meaningful for others only within these cultural frameworks."[27]

Each chapter takes up the task of thinking about an extinction event through intertwining conceptual and ethical demands while inquiring into the cultural forms and media that have been used to represent the loss of an entire life form. I do not present a survey of all the extinction events in the past few centuries and the debates over causes and remedies. Instead, I focus on a small sample of events in which ideas of extinction underwent extreme intensification and redefinition. Reckoning with these extinction events involves examining what happens when the philosophical, psychological, and cultural forms used to account for species finitude are pressed to their limit ends, as well.

Accordingly, this book examines extinction as a concept and the extinction of concepts themselves. The terms and disciplines used to write about extinction also are caught up in situations of their own crisis, thrust into a contested condition of "lastness" of their own. Yet the scene of the end of writing arises in different historical contexts that point to disputed outcomes for encountering the end of the imagination. In some cases, narrating the demise of the literary has resulted, paradoxically, in initiating new kinds of creativity and legibility regarding shared ecological precarities. In other cases, the double finitudes of life and art coincide with apocalyptic perceptions and claims for the expedience of asserting nationalistic, anthropocentric, survivalist, and salvationist agendas. In all cases, these scenes of extinction are instances of the extremes of address, an address to the end of the addressability of a species. The case studies presented here inquire across different historical junctures who, if anyone, is called to read these limit ends, what kinds of readers congregate around these scenes at specific junctures, and what happens to reading itself in such moments?

## "And so here I am left alone": On Witnessing One's Own Extinction

This book builds on a number of recent studies of extinction that have begun to establish a broader scope for how to understand the multiple meanings of extinction across the sciences and humanities. Work by Deborah Bird Rose, Ursula Heise, Claire Colebrook, Ray Brassier, Thom van Dooren, Matthew Chrulew, Brent Buchanan, Audra Mitchell, Mark V. Barrow Jr., Ashley Dawson, Susan McHugh, and Elizabeth Kolbert, among others, has helped establish a broader understanding of the language and ideas used to make sense of a variety of extinction events and the cultural responses to such existential ends.[28] These scholars have provided powerful reflections on how the understanding of extinction in present times continues to shift and require multiple perspectives and methods.

Within the field of extinction studies in the humanities, there has been a divergence in works that focus on the conceptual problems that follow in the wake of knowledge of species death but hardly mention actual animals (Brassier) and those that emphasize detailed study of the specific lives of endangered or extinct animals but only briefly touch on questions of the broader philosophical implications of extinction (Kolbert). Another divergent tendency exists between works that examine nonhuman animal extinction only (without reference to the history of human and genocidal extinction events) and works that emphasize primarily the effect extinction has on humanity. The term *extinctions*, in the plural, carries the tremendous weight of all these different kinds of extinction events, and its use and history must be parsed carefully across the human–animal condition. This book establishes an integrated and interdisciplinary approach in analyzing the history and unique consequences of each instance of species loss together with a reflection on the connected impact of concepts of extinction across human and nonhuman animal life.

One of the aims of this book is to examine how and why different ways of representing and responding to extinction appeared at specific historical moments and in particular locations at the nexus of culture and biogeography. In 1863, the Reverend Charles Kingsley published the first example of a new, unique narrative perspective: an account of extinction told from the point of view of the last of the species; in this case, the great auk. Kingsley, an early supporter of Darwin, placed a short, first-person account of the last great auk in his children's novel *The Water Babies*.

The great auk, a three-foot-tall flightless bird in the genus *Pinguinus* (great auks are not directly related to penguins, although those birds were named after the great auk due to similarities in appearance), dwelled on a number of scattered islands in the North Atlantic from Canada to the United Kingdom. An important food and cultural symbol to coastal Indigenous Peoples of the North Atlantic, the bird and its eggs were frequently hunted by European and settler North American fishers and sailors. Hunting of the great auk increased significantly in the early 1800s as maritime trade and commercial fishing rapidly expanded. By the 1830s, a time when the first arguments for the natural science of extinction had become widespread, sympathizers of the great auk already felt the bird was not likely to survive. The last recorded witnessing of a live animal occurred in 1852.

A decade later, Kingsley had the idea to have the last great auk tell its story in a children's fantasy novel. In the spirit of the novel's whimsical style, the solitary great auk has a flair for levity while telling of its last days in "Allalonestone," a fictional island off Scotland. Kingsley uses the great auk's Icelandic name Gairfowl and describes the last female of the species as "A very grand old lady, full three feet height, and bolt upright, like some old Highland chieftainess." Here is her tale:

> We have quite gone down in the world, my dear, and have nothing left but our honour. And I am the last of my family. A friend of mine and I came and settled on this rock when we were young, to be out of the way of low people. Once we were a great nation, and spread over all the Northern Isles. But men shot us so, and knocked us on the

head, and took our eggs—why, if you will believe it, they say that on the coast of Labrador the sailors used to lay a plank from the rock on board the thing called their ship, and drive us along the plank by hundreds, till we tumbled down into the ship's waist in heaps; and then, I suppose, they ate us, the nasty fellows! Well—but—what was I saying? At last, there were none of us left, except on the old Gairfowl-skerry, just off the Iceland coast, up which no man could climb. Even there we had no peace; for one day, when I was quite a young girl, the land rocked, and the sea boiled, and the sky grew dark, and all the air was filled with smoke and dust, and down tumbled the old Gairfowl-skerry into the sea. The dovekies and marrocks, of course, all flew away; but we were too proud to do that. Some of us were dashed to pieces, and some drowned; and those who were left got away to Eldey, and the dovekies tell me that they are all dead now, and that another Gairfowlskerry has risen out of the sea close to the old one, but that it is such a poor flat place that it is not safe to live on: and so here I am left alone.[29]

Kingsley presents the last female auk as a plaintive storyteller of her own species' demise from her point of view, divulging character traits of both dignity and shame in delivering this last discourse. The narrative mixes fiction and nonfiction, as a volcanic eruption in Iceland in 1830 did have a devastating effect on a small remaining population of great auks. Every sentence of the garrulous Gairfowl carries the weight of the declaration "I am extinct," an uncanny speech act akin to the utterance "I am dead" that Jacques Derrida identifies as the paradoxical conjunction of possibility and impossibility in a sentence that can never be spoken literally in the first person.[30] Extinction is a magnitude well beyond individual death, as the last animal carries the extreme burden of being the last speaker ever for its own species. No one can ever say "I am extinct," yet the phrase is a possibility, indeed an inevitability, definitive of belonging to a species.

The last great auk goes on to tell the story of the second to last great auk, a "gentleman [who] came hither with me." She continues:

After we had been here some time, he wanted to marry—in fact, he actually proposed to me. Well, I can't blame him; I was young, and very handsome then, I don't deny: but you see, I could not hear of such a thing, because he was my deceased sister's husband, you see? . . . I felt it my duty to snub him, and howk him, and peck him continually, to keep him at his proper distance; and, to tell the truth, I once pecked him a little too hard, poor fellow, and he tumbled backwards off the rock, and—really, it was very unfortunate, but it was not my fault—a shark coming by saw him slapping, and snapped him up. And since then I have lived all alone (182–83).

In this first self-elegy written by an extinct animal,[31] the display of sorrow and isolation with touches of tragicomedy resonates beyond the frame of fiction. The speaker presents the persona of an old maid figure whose sexual crises are tantamount to her species' crises. In her reflections on her experiences of youthful desires, human violence, and the misadventures of her fellow final species, she maintains a sturdy sense of family honor and something akin to British fortitude. Kingsley solicits a sense of regret for the great auk, but he does not think this emotion need last. The narrator reflects that "the old Gairfowl is gone already: but there are better things to come in her place" (184). Those who would now come to the island would find "eighty miles of codbank, and food for all the poor folk in the land. . . . And then we shall not be sorry because we cannot get a Gairfowl to stuff" (184). However, the cod population would undergo its own collapse in the 1980s due to overfishing, and many cod communities today are designated as threatened.

This short narrative of the last great auk already establishes a number of prominent themes that will appear across many accounts of last animals; that is, animals facing extremely low population numbers, approaching zero. Last animals often are heavily anthropomorphized and cast into a typological character (for example, wise elder, lost soul, or melancholic icon) that presents them as spokespersons for their species. The role of humans in the population collapse is mentioned as the proximate cause of the auk's decline, but the last animal spokesperson does not spend much time castigating humans. Rather, there is already a hint of resignation, loss of sexual appetite, and acceptance of lastness, whereby the final great auk's individual persona becomes framed as species character traits.

A few decades before the bird's disappearance, the British naturalist Thomas Nuttall, in his *Manual of the Ornithology of the United States and Canada, 1832–34*, insisted that the great auk had dark and antisocial characteristics that contributed to its impending downfall. "Degraded as it were from the feathered rank, and almost numbered with the amphibious monsters of the deep, the Auk seems condemned to dwell alone in those desolate and forsaken regions of the Earth."[32] Here, the animal is depicted as physically monstrous and seemingly condemned to a life alone. In these early extinction narratives, as animals approach last numbers, they often become objects of pronounced moralization and blame (the dodo supposedly is named for its stupidity). For Kingsley, it would seem the proper ethic for this last animal is to die well, to perish with dignity, and depart with not too much complaining—illustrating a lesson for Victorian-era children to admire such animals but not get too attached to their lives and deaths.

This book studies the development of both factual and fictional accounts of species finitude, and though I am careful not to conflate imaginary stories of animal loss with the documentation of actual animal deaths, I discuss how the understanding of extinction has required an intertwining of natural and cultural histories that entangle factual and interpretive responses to species mortality. Every extinction calls for a forensic question: What happens when a species approaches extremely low numbers? Other questions follow immediately: Why did this particular species go extinct? How does a specific extinction event affect definitions and attitudes toward life and death? How has the natural history of extinction changed over the past several hundred years? What are the cultural forms and debates that come to be called upon in the situations of last animals?

These questions call upon the title of this book, *What Is Extinction?*, a question that persists throughout this study and cannot be answered once and for all. One might simply say that extinction is when the population of a species is zero. Yet this definition is complicated by debates on the ambiguities of species, subspecies, hybridization, and variation classifications in the natural sciences by biologists as far back as Darwin. Moreover, we want to know what extinction is in a comprehensive way as more than just a number. A single species extinction is a zero and a finality; but extinction is also an event with multiple natural and cultural repercussions and an extended process whose definitions have changed over time and continue to change today.

## On the Origins of the End of Species

The emergence of a scientific understanding of extinction based on empirical evidence from the analysis of fossils (methodologically distinct from religious notions of apocalypse that ascribe theological causality to the finitude of life[33]) dates to the late eighteenth century. The proving of extinction using methods of comparative anatomy applied to fossil evidence marked a high achievement for Enlightenment science. Yet, while Enlightenment thinkers sought to link the development of natural scientific knowledge with both species complexification and civilizational development, the revelation of extinction presented a scenario of intellectual progression as devastation. Extinction forced these thinkers to contend with how the devastating loss of species and species flourishing could be effectuated in the same process. The breakdown of species life could not be extricated from the building up of species life. Naturalist scientists found that at the heart of the study of life was the study of death. Extinction required a recognition that enlightenment and endarkenment, or evolutionary development and catastrophe, were not mutually exclusive.

How did the scientific determination of extinction become possible in the first place? Theories of species change already had appeared by the middle of the eighteenth century, when initial insights into extinction were developed. The argument that species altered over time paired with the assertion that they could disappear altogether overturned the two previously dominant discourses on life, Aristotelianism and Christianity. Aristotle maintained Plato's assertion of the "principle of plenitude" in nature (in Arthur Lovejoy's phrasing)[34] arguing that the totality of life had already been created. Aristotle also added a "principle of continuity" (58) insisting that all entities fell under causal laws, and that "nature does nothing in vain."[35] Furthermore, Aristotle claimed that "nature never creates anything without a purpose, but always what is best in view of the possibilities allowed by the essence of each kind of animal."[36] Since everything in nature exists in accordance with the overall purpose and causality inherent in the natural order, Aristotle could not have conceived of an animal that wholly disappeared, since such an animal would have a cause that would be to no purposeful end.

Medieval Christian scholars used Aristotle's natural history as the basis for constructing a hierarchical *scala naturae*, with humans on top. In this medieval paradigm, as John G. T. Anderson summarizes, "The *scala* was governed by three principles: plenitude, continuity, and gradation."[37] God created all possible species, there were no gaps or broken links in the chain of species, and all life was hierarchically organized. When presented with evidence of fossils, church theologians in the thirteenth century claimed them as "models of God's rejected works" or "outlines of future creations."[38] For Aristotle and the medieval Christian naturalist theologians, everything had a purpose in nature and every species had a place. Under these conditions, extinction was a conceptual impossibility.

How, then, did it become conceivable that a species could be effaced permanently from historical existence, puncturing a hole in nature? The rapid rise of the natural sciences in the seventeenth century followed on the expansion of scientific methods based on empirical observation, inferential reason, and a preference for the mechanistic explanation of phenomena. By the eighteenth century, the rush was on to decode nature into a series of natural laws and detail these laws in encyclopedic fashion. One of the more remarkable attempts at organizing the entirety of natural history can be found in the work of George-Louis Leclerc, later named as Comte de Buffon. As head of the Royal Botanical Gardens of France, with access to one of the largest natural history collections in the world, the massive "Cabinet of the King" (later to become the primary resource for the Muséum National d'Histoire Naturelle in Paris), Buffon set himself the task of nothing less than a total explanation of the natural world with his *Histoire naturelle, générale et particulière*, first published in 1749 and revised and expanded upon until his death in 1788. He developed a materialist and historicist approach to nature, declaring that "Natural history taken in its full extent, is an immense History, embracing all the objects that the universe presents to us. This prodigious multitude of Quadrupeds, Birds, Fish, Insects, Plants, Minerals, etc., offers a vast spectacle to the curiosity of the human spirit."[39]

Translating this curiosity into a sweeping panoramic vision of natural history, Buffon found a place for everything, from minute anatomical data of individual animals to speculations on the origin of the universe. Critical of Carl Linnaeus's ahistorical and timeless taxonomies, Buffon declared: "Nature, I admit, is in a continuous movement of flux,"[40] and "the great worker of Nature is Time" (307). Buffon began to craft theories of the formation of the Earth and the generation of species as processes working through temporal change rather than being eternally static or fixed according to pre-set teleological ends. Applying this thinking to the temporal development of species, Buffon claimed that some animals "often changed to the point of being unrecognizable" (307).

The same argument for species change could be extended to account for a species that might fail to change in order to survive the temporal flux. Buffon pointed to the giant "incognitum" of North America, later called the mastodon, whose skeletal remains were unearthed in the mid-eighteenth century. In Buffon's view, the absence of any living evidence of the massive animal was more easily explained than speculations on the possibility of its survival in some remote area of the continent. The mastodon, according to Buffon, "was certainly the foremost, the largest, the strongest of all quadrupeds: since it has disappeared, how many others—smaller, weaker, and less noticeable—must have perished without having left us either evidence or information about their past existence?" (307).

If the powerful, dominating mastodon failed to survive despite its strength, Buffon reasoned that countless other species that did not have these advantages must have succumbed. Buffon then proposed the general claim that, in some cases, species can be "erased": "Everything reminds us of these monsters by their failure, these imperfect rough drafts, a thousand times tried and executed by Nature, which, having barely the ability to exist, must only have survived for a time, and have since been erased from the list of creatures" (308). Arguing that the effects of temporality had to apply to natural history as well as to the methodology used to study it, Buffon enjoined naturalists to "glance behind and ahead, to try to catch a glimpse of what [Nature] could have been before, and what she may become in the future" (308). Doing so would open considerations that "the least perfect, most delicate, heaviest, least active, least armed, etc. species have already disappeared or will disappear" (307). Expanding the age of the Earth backward and forward suddenly made it conceivable that the appearance and disappearance of species had a natural history. The concept of extinction, previously unthinkable, now suddenly seemed inevitable.

Buffon was not the first naturalist with the inkling that an entire species might disappear,[41] but he offered the first comprehensive theory of natural history that included an explanation of extinction as a fundamental aspect of speciation. Buffon's argument for extinction was only a minor part of his overall oeuvre, but one of his successors, Georges Cuvier, established his career as a naturalist in large part on elevating extinction into a robust natural science.

Cuvier's first major scientific paper, published in 1796, demonstrated his formidable skills in comparative anatomy by showing that the teeth and jaw shape of mammoth fossils found in Europe and Siberia did not match those of living elephants. Cuvier did not have a ready explanation for the processes involved in extinction, but he insisted the question of the disappearance of a species needed to become central to natural science. "What has become of these two enormous animals [mammoth and mastodon] of which one no longer finds any traces, and so many others of which the remains are found everywhere on earth of which perhaps none still exist?"[42]

Cuvier assumed that only some sort of planet-wide violent, catastrophic event could explain the disappearance of a whole species, a thesis he would maintain all his life. In his first paper, he set naturalist minds racing as to what this catastrophe could be and how one might imagine it. "All these facts, consistent among themselves, and not opposed by report, seem to me to prove the existence of a world previous to ours, destroyed by some kind of catastrophe, beings whose place has been filled by those that exist today, which will perhaps one day find themselves likewise destroyed and replaced by others. But what was this primitive earth?" (24). In subsequent papers, Cuvier postulated that catastrophes set off new geological epochs, and he showed how extinctions in different locations across the Earth could be dated to the same geological periods.

The call to think that catastrophe, the destruction of animal life, and the "primitive earth" involved investigating what Cuvier described as the Earth's "revolutions,"[43] applying the political term for dethroning the monarchy to the upheaval of species. Cuvier continued his research with a subsequent paper in which he itemized evidence of extinction for twenty-three species "all quite certainly unknown today, and which all appear to have been destroyed, but whose existence in remote centuries is attested by their remains" (53). At the same time as he meticulously used a knowledge of comparative anatomy to reconstruct lost species, Cuvier stoked public and scientific interest in speculation on the enigma of the natural history of destruction. "How can one repress such a natural desire to give an account of causes that have been able to produce such terrible effects: to raise mountains, to shift seas, to destroy whole species, in a word to change the face of the globe and the nature of the beings that inhabit it?" (40).

Such questions brought together visions of the geological sublime and the romantic appeal of sifting through ruins with the studious determination to gather empirical evidence from fossils and rock strata. Cuvier sought to collect as many fossils from around the globe as he could, while he also collaborated with the geologist Alexandre Brongniart on a study of the geological layers of the Paris basin, which had been exposed by open pit mines outside of Paris. By locating fossils of extinct animals in specific strata and then comparing fossils found in other strata that Cuvier assumed formed over "thousands of centuries" (70), this research prompted the need to think in much longer time scales than previously had been proposed. While Cuvier's writings and reconstructions of lost species elicited speculations on the age of the Earth on a much more grand and ancient time scale than the proposals of other natural historians, this history also cast the Earth into multiple abysses and devastations. Cuvier's work opened a breach in naturalist thought by inaugurating a science of the end of science in the comparative study of the ends of lifeworlds. Yet in a retreat from these destabilizing and disturbing views of lifeworlds undone, Cuvier also came to apply these same methods of comparative anatomy toward bolstering a science of racial hierarchy, one that asserted white European supremacy.[44]

Michel Foucault, in *The Order of Things*, provocatively argues that "there is no life"[45] in the natural history worldviews preceding Buffon because previous natural scientists did not grasp life as an independent conceptual problem that stood apart from presuppositions of a timeless ontological order. Only toward the end of the eighteenth century, in Foucault's assessment, did naturalists begin to formulate biological and evolutionary thought on "life itself" with the introduction of methods attuned to the temporalities of historical change. Foucault effusively credits Cuvier with laying the foundations for a science of biology with his development of comparative anatomy.

For all his praise of Cuvier, however, Foucault, oddly, does not mention at all Cuvier's research on extinction that Cuvier had paired throughout his career with investigations into comparative skeletal and morphological structures (Foucault uses the term "extinction" only once in passing in the book, a difficult term to miss when commenting on Cuvier. Foucault also does not discuss Cuvier's racist writings on human evolution). But Foucault does observe that burgeoning redefinitions of life and death marked an emerging "analytic of finitude" (344) in Cuvier's time. In Foucault's assessment, Cuvier's insights divulged the worldview that "the animal appears as the bearer of that death to which it is, at the same time, subjected; it contains a perpetual devouring of life by life. It belongs to nature only at the price of containing within itself a nucleus of antinature" (302).

Foucault's "antinature" is the work of finitude that has, paradoxically, contributed to the structuring of nature. Foucault also calls this an "untamed ontology" (303) whereby species are understood to emerge, flourish, and perish in a condition in which all species share the same contingent and temporal situation, absent of any teleological or "tamed" ontology that guarantees survival. However, his phrase "a perpetual devouring of life by life" does not accurately describe extinction (nor, for that matter, co-adaptation), and it suggests a continual process pitched at the theoretical construct of "life itself" rather than the geographically specific existences of species that are susceptible to disruptions and terminations.

Even though, by the end of his study, Foucault will invoke the idea of the disappearance of the organizing figure of the human generated in the nineteenth century, washing away "like a face drawn in sand at the edge of the sea" (422), he still is not referring to extinction. Instead, Foucault posits that the dissolution of the figure of the human would mark the end of the "episteme" of finitudes and historicities that furnished the basis for the human sciences and constituted the modern figure of "man." This washing away of epistemes profoundly sets the stage for redefining humanity and life itself as having a future existence after "man" (imagining, perhaps, something akin to Nietzsche's superman). However, the episteme of extinction—which contributed to the birth of a non-anthropocentric ecological paradigm comprised of species precarities and entanglements that do not promise any future horizon for humanity—did not shape Foucault's thinking.

Cuvier's evidence for lost species posed the challenge of how to imagine past worlds while also conceiving of the catastrophic ends of these worlds at the same time. This insight into the alternative, forlorn worlds embedded in the Earth's crust tempted readers of Cuvier into viewing the entire planet as a romantic ruin. In Honoré de Balzac's novel *La Peau de chagrin* (1831), the narrator proposes, "Is not Cuvier the greatest poet of our age?"[46] Balzac saw in Cuvier's research that powers of the mind could become analogous to the powers of geological processes at work over eons. With Cuvier as guide, one could now glimpse how inert stones could tell the story of animal worlds, and how numerous lost worlds lay buried dormant under foot.

> Have you ever plunged into the immensity of time and space by reading the geological tracts of Cuvier? . . . As one penetrates from seam to seam, from stratum to stratum and discovers, deep in the quarries of Montmartre or the schists of the Urals those animals whose fossilized remains belong to antediluvian civilizations, the mind is terrified to catch a vista of the billions of years and the millions of people which the feeble memory of man and an indestructible divine tradition have forgotten. . . . He excavates a fragment of gypsum, spies a footprint and shouts: "Behold!" And suddenly marble turns into animals, dead things live anew, and lost worlds are unfolded before us! (19, translation modified).

However, Cuvier's catastrophist theory of geological history soon was challenged by Charles Lyell's "uniformitarian" argument that claimed extinction was regular and recurrent, consistent with the slow-churning geological processes. In the second volume of *Principles of Geology* (1832), Lyell insisted that extinction was not rare or caused by apocalyptic events but, rather, predictable due to the regular laws and motions of earth systems. Lyell also considered that the rise of one species "must in like manner, have marked its progress by the diminution, or the entire extirpation, of some other."[47]

For Lyell, extinction should be expected given the "progress" of species population dynamics. Lyell was quick to tie this observation with an analogy supporting colonialism, as he added: "We must at once be convinced, that the annihilation of a multitude of species has already been effected, and will continue to go on hereafter, in certain regions, in a still more rapid ratio, as the colonies of highly-civilized nations spread themselves over unoccupied lands" (276). Lyell understood that colonial conquest of "highly-civilized nations" procuring "unoccupied lands" would result in further extinctions, but he viewed this anthropogenically-caused destruction of species as par for the course.

There is a remarkable footnote in Lyell's tome on geology where he briefly brackets scientific rigor and divulges a sentimental anguish when reflecting on the magnitude of extinction. Lyell describes himself pondering the decomposition of the remains of a dodo in a natural history museum, and wonders what kind of grave would be adequate to such loss.

> Some have complained that inscriptions on tomb-stones convey no general information except that individuals were born and died, accidents which must happen to all men. But the death of a *species* is so remarkable an event in natural history, that it deserves commemoration, and it is with no small interest that we learn, from the archives of the University of Oxford, the exact day and year when the remains of the last specimen of the dodo, which had rotted in the Ashmolean museum, were cast away (273).

That date was January 8, 1755, which Lyell observes on a note left in Latin by a museum curator. In Lyell's view, extinction required not only a scientific explanation but also a special elegiac appreciation. Lyell remarks on the double loss of the last dodo: first, as it became extinct, and second, as its rotted remains were discarded, leaving behind only a note as to the date when the remnants were swept away. For Lyell, the death of species had to be understood as a remarkable and traumatic event in natural history, one that caused a special kind of shudder to ripple in the living observer.

By the time Darwin wrote in *On the Origin of Species* that "No one I think can have marvelled more at the extinction of species, than I have done,"[48] the science of extinction had reached a point of maturation. Darwin actually guided his readers to downplay astonishment at the evidence for extinction while insisting, at the same time, that no one could out-marvel him in his determination to resolve the problem of how to combine a theory of speciation with a theory of extinction. Understanding both phenomena required a theory of evolution that took a very long view on the vicissitudes of species. From such a perspective, extinction appears as a regular occurrence and not indicative of a planet prone to catastrophes. "So profound is our ignorance, and so high our presumption, that we marvel when we hear of the extinction of an organic being; and as we do not see the cause, we invoke cataclysms to desolate the world, or invent laws on the duration of the forms of life!" (61). Unlike Lyell, Darwin did not extend any special grief toward any particular lost species. In Darwin's view, profound marvel, instead, was warranted in considerations of all the sundry intricacies involved in sustaining any particular life form. "We need not marvel at extinction; if we must marvel, let it be at our presumption in imagining for a moment that we understand the many complex contingencies, on which the existence of each species depends" (260). Darwin not only rejected Cuvier's catastrophism as the single causal explanation for extinctions, he also refused to see extinction as having catastrophic effects on those who think what it has wrought.

Refusing the rhetoric of disaster or endarkenment, Darwin insisted that extinction could be expected when a species' population numbers dwindled due to a variety of "injurious agencies."

> It is most difficult always to remember that the increase of every living being is constantly being checked by unperceived injurious agencies; and that these same unperceived agencies are amply sufficient to cause rarity, and finally extinction. We see in many cases in the more recent tertiary formations, that rarity precedes extinction; and we know that this has been the progress of events with those animals which have been exterminated, either locally or wholly, through man's agency (258).

Given the omnipresence of lethal scenarios, Darwin insisted that extinction should not be considered an unusual outcome. Darwin's advocacy for a staid, nondramatic study of species ends serves as a kind of consolation for thinking about extinction. He viewed the origins and ends of species not as fixed markers but as momentary shifts and terminations in an overall process of the "economy of nature" (260). For Darwin, the conditions of possibility for speciation were intricately intertwined with the conditions of the end of possibilities for other species. "On the theory of natural selection the extinction of old forms and the production of new and improved forms are intimately connected together" (256). Yet in ultra-catastrophic conditions such as mass extinction, this chain of "the production of new and improved forms" can break down. One can transpose this argument into a Hegelian language: extinction is a negation that can, in turn, be negated (sublated) and become the spur for a new phase of speciation. However, extinction also can be an event of negativity that cannot be recuperated, an utterly undialectical and radical loss of a species form, which builds nothing but only leaves a tear in the fabric of reality.

Although Darwin insisted on muting any surprise or astonishment over the revelation of extinction, his work had the opposite effect. He brought the awareness of the reality of extinction to a wider audience and made reckoning the end of any species, including humans, a palpable and present problem. If speciation provoked naturalists like Darwin to express effusively a sense of wonder and enchantment at the variety of life, the end of a species also stirred a pained, somber form of wonder caught up in ghostly figures and perceptions of eradicated life. These visions of animal loss were stunning on their own, but this knowledge of radical finitude also began to change the pursuit of scientific knowledge more generally. Extinction required the study of the annihilation of one's object of study, a scientific method for the demise of the scientific method. Research in geology, natural history, and astronomy began not only to show evidence of worlds thriving without humans but also the impermanence and extinguishability of all lifeworlds.

The astronomer William Herschel first began to conceptualize the birth and death of nebulae (later ascertained as galaxies) by the end of the eighteenth century. Herschel extrapolated his observations to Earth's position in the universe and came to the conclusion that the Milky Way cannot last forever.[49] A few decades later, William Whewell, a naturalist polymath and near-contemporary of Darwin, wrote that the new sciences of biology, geology, and astronomy accounted for the transience of humans as well as every other object in the universe. Whewell declared: "Not only the rocks and the mountains, but the sun and the moon have the sentence 'to end' stamped upon their foreheads."[50] In the heart of the Enlightenment, extinction showed the way toward the end of the Enlightenment. Instead of a narrative of species' progress and improvement, extinction threw open the gates toward unknown outcomes for any form of life, while indicating that all living forms would eventually transform into ruin. The question "What is extinction?" suddenly became an inquiry that could question everything.

Balzac, just after his fictional praise of Cuvier as the greatest poet, lamented upon what was left to live and write after such knowledge of the long history of lost lives. "We wonder, overwhelmed as we are by the destruction of so many past universes, what use are our glories, our hates, our loves; we wonder whether it is worthwhile to accept the pain of life in order that hereafter we may become an intangible speck?"[51] Geological strata and fossils suddenly became legible and vast scales of geological time became cognizable, but at the cost of radically questioning the purpose of life lived at human scale. At the same time as the geological and biological wonders of the planet became readable, one could see the world already as on the path of destruction, in which yet another "revolution" would efface the current lifeworld. The initial insights into the biology of extinction had effects well beyond the formulation of a scientific consensus on the finality of species loss. As Balzac attests, the thinker of extinction is implicated in the thought of the end of thought, an epiphany both thrilling and dreadful, opening up sweeping vistas for the imagination while at the same time compelling works of culture into beholding the future ghost of life itself.

## On the Origins and Ends of Species

To understand the broad ramifications of extinction today, it is necessary to have a coherent yet flexible concept of the species form. Any theory of how life is imbricated with extinction must address "the species problem": namely, it is not clear that we even know how to define species.[52] Are species a natural kind or a classificatory convention? Should species be defined by DNA, descent, shared capacity to sexually reproduce, structural homology, regional and temporal isolation, or some other criteria? How do we differentiate between species and the process of speciation?

In several instances in *On the Origin of Species*, Darwin recognizes the ambiguity of the designation "species," and even welcomes the conceptual vagueness that comes with the term. "Nor shall I discuss the various definitions which have been given of the term 'species.' No one definition has satisfied all naturalists; yet every naturalist knows vaguely what he means when he speaks of a species."[53] Darwin repeatedly states that he finds no consistent way to distinguish species from varieties, but the ambiguity of the concept of species does not get in the way of his investigations. Rather, he is able to better theorize speciation because he does not insist on a strict definition of species even while he retains the species form as important to biological processes. He keeps the species form even as he deconstructs it. The looseness of the species category distances his ideas from rigid essentialism or previously fixed and static taxonomies.

Darwin's arguments stand in contrast to more recent biological reductionist claims such as those made by Richard Dawkins that it is the gene pool and the genotype, rather than the species form or phenotype, that is the unit of biological selection. For Dawkins, the loss of a species form is not all that important, as long as some material of the genetic code or gene pool continues on in other species. In Dawkins's view, one of the gene's defining characteristics is its apparently endless capacity for regeneration. In *The Selfish Gene*, Dawkins mentions that he could have titled the work *The Immortal Gene* on the suggestion of a friend and calls DNA "immortal coils."[54] In this model, the extinction of a species form is a minor disruption in the ongoing and continual replication of various genomes.

Dawkins's theory of the "immortal" and reproductive replicating genome draws his apparently reductionist claims closer to metaphysical theories of the continual becoming of life. Even though Dawkins would claim nothing but scorn for vitalism or panpsychism—metaphysical philosophies based on the notion that life and consciousness are cosmically present in all matter in some rudimentary form, a claim that is wholly incompatible with extinction since it posits life as permanently inherent to substance—his theory follows similar lines by venerating the endlessly replicating gene and downplaying the importance of extinction as the loss of a species form.

Thinking about the ramifications of extinction requires recognizing the species form as the manifestation of the intertwining of genotypes and phenotypes formed in symbiotic as well as auto-immune activities. The species form can be described as a *moving baseline* that indicates the integrity of the species form even as genotypes and phenotypes can fluctuate. A moving baseline allows one to track how a species can change or adapt in more than one dimension, or how environmental disturbances might overwhelm the possibility for a species to change. This moving baseline is not reducible to a nominalism; nor is it heuristic. Rather, it accounts for the shifting qualities of embodiment experienced by a species, including its local symbioses and co-adaptations, while respecting how the extinction of a unique life form effaces both genotype and phenotype.

In this book, I do not offer a new theory of "life itself" that is undone by "death itself," nor am I preoccupied with discerning the exact physical nature of life and death or the dividing line between the organic and inorganic. I also do not provide an analysis of how concepts of extinction change at the level of microbial life, a form of life without individuality, for which the fields of microbiology and epidemiology employ definitions of life and death unique to these disciplines. Instead, my analysis builds on interdisciplinary interpretations across biology, ecology, and the humanities of well-known animal and human extinction events that have become widespread public concerns and contributed to fundamental redefinitions of the life and death of species. The moving baseline conceptualization of the species form is meant to describe both the natural and cultural ways (or what Donna Haraway calls the "material-semiotic"[55] conditions) of discerning and understanding species.

However, the history of species categorizations is rife with political acts of domination, from long-standing and ongoing colonialist extractivism of Indigenous lands and lives used to furnish Western academic research to the epistemic condescension of Latinate systems of naming that displace local and vernacular knowledges. With careful attention to the ethics of naming and identifying life, it still remains important to remark that one cannot comprehend the severity of extinction without also thinking of the uniqueness and permanent loss of the species form.[56]

Extinction can entail a partial or full loss of some genes from a gene pool, but it also is the total loss of a species presence and way of life, which will never be repeated and will no longer have effects on shaping environments and cultures. Eileen Crist describes such loss as "annihilating a living cornucopia that is self-replenishing and self-creative."[57] Furthermore, this moving and creating baseline of the species form is necessary to provide the concept of biodiversity with its own integrity in order to understand it as something variously embodied and experienced rather than merely equated with statistical gene pools.[58] Species can continue to change or become modified until the event of extinction, which is the only point at which the "essence" of a species (its genotype and phenotype) is truly fixed. This argument for a dynamic conception of the species form follows multispecies paradigms of interconnected life while also attending to the singularity of each species.

While reductionist and vitalist philosophies tend to downplay or dismiss altogether the importance of extinction and its effects on the species form, a very different kind of assessment of the centrality of extinction can be found in the philosophical writings of Ray Brassier. In *Nihil Unbound: Enlightenment and Extinction*, Brassier argues that extinction discloses a fundamental reality that is devastating for both life and thought: namely, that there is no larger horizon of meaning for life or Being, and that the eventual extinguishing of human thought annuls any philosophy of transcendental subjectivity. Brassier's book is a dense and intricate work that considers how the "transcendental trauma"[59] of extinction undoes much of the claims continental philosophers have been making for the past few centuries regarding the question of the meaningfulness of being and the phenomenological coordinates of subjectivity. Instead of elevating the subject to a figure whose phenomenal experience and epistemic coordinates are the gateway to the significative capacity of Being, the fact of extinction forces the subject to think about its own purposeless ends in a universe indifferent to the dissolution of life.

Brassier's writings merit further examination here since his work advances the most cogent argument that extinction goes well beyond the facts of biology and ecology and fundamentally reshapes the entire philosophical project around the disenchantment of nature as "home" or any claim that "life itself" or the apotheosis of subjectivity shows the pathway toward meaning in the universe. The truth of extinction, Brassier insists, reveals the internal limits of mind, world, and sense to an external and, ultimately, cold, objective, nonconscious, non-meaningful universe. "Extinction turns thinking inside out, objectifying it as a perishable thing in the world like any other. . . . This is an externalization that cannot be appropriated by thought—not because it harbours some sort of transcendence that defies rational comprehension, but, on the contrary, because it indexes the autonomy of the object in its capacity to transform thought itself into a thing."[60] Brassier's argument follows along the lines of philosophical naturalism and eliminativist materialism, which insists that the subjective qualities of life can be explained by objective biological and chemical processes. Thus, there is no reason to assume any special metaphysical status to life or its ends.

This commitment to thinking about the ramifications of extinction does not compel Brassier to assume a particularly apocalyptic outlook, and he maintains that scientific realism and the pursuit of collective social justice remain as valid norms and objectives. While I do not pursue here the nuances of Brassier's arguments for what the extinction of thought entails for philosophy more generally,[61] I will note that Brassier's insights are pitched at a conceptual level that does not sufficiently clarify immediate ecological concerns. By viewing extinction at the level of a "transcendental trauma," Brassier subsumes Earth's biological extinctions within a universal, cosmic extinction that is the ultimate factual reality of space and time. However, why should the time of extinction of all life supersede any other concept or experience of time, including the chronological, the immediate, or the proleptic anticipation of death in the being and time of existential thought? Even if extinction is inevitable, does that make all other forms of time collapse into one objective "time of death" (161)? There may be more than one temporal frame in question for any being, and the sequential passage of time cannot be collapsed into one end time. The universe must pass through temporal stages, and even if these temporalities are all extinguishable, we cannot skip these and just jump to a generalized extinction.

Brassier's inquiry does not delve into the details of how extinction events happen and how the specific stages of reaching a zero point for any particular species has both a biological and philosophical import. He also finds little relevance in the species form for thinking about extinction, favoring cognitive crises over ecological ones. How can we understand the extinction of biological life on Earth as intertwined with but still distinct from the epochal, "transcendental efficacy" (230) of cosmic extinction that is the "anterior posteriority" (230) that foreordains the annihilation of all life?

This question of how to combine a deep-time vision of total extinction with current trends of increasing rates of species loss is connected to a practical ethical dilemma in our own time: even if extinction is the eventual reality facing all species, this does not let us off the hook right now to just wipe out the biodiversity on the planet for our own immediate gratifications. There are at least two temporal realities to species extinction: the current rapid loss of species and the inevitable futural loss of all species. How should we think and act upon these together? How might we maintain an ecological thought of biological extinction on Earth and a philosophical thought of transcendental, cosmic extinction as intertwined but still distinct processes?

If everything is alive, as in vitalism or immortal genes, then extinction does not matter. If everything is dead already, as in a version of eliminativist materialism, then it is a real question if anything can be said to matter. Thinking about extinction entails taking the objective, non-meaningful or "nihilist" conditions of nature seriously yet also taking the current contingent conditions of life seriously. Extinction is anti-foundational for philosophies of life and theories of transcendental subjectivity but foundational for ecological thought. "Transcendental trauma" makes no sense if one is entirely neutral to the difference between universal dissolution and actually existing ecological states with their unique concatenations of species. Without loss and extinction, as in philosophies of endless becoming, there is no ecology. But with too much loss and extinction, there is, also, no ecology.

## Toward an Ecological Theory of Extinction

This book affirms the thesis that extinction is a central factor in making ecology coherent in the first place, but extinction also presents a number of crises and contradictions for ecological thought. While the meanings and terms associated with extinction have shifted a number of times over the past few centuries, I want to articulate some of the broader challenges that extinction poses for conceptualizing ecology and ecological art. Presented here are a series of central themes that form the basis of this book. They concern how extinction has become knowable and how the consequences of such knowledge can be both powerful and unsettling to the thinker.

*Thinking extinction implicates ecological thought in a number of paradoxes.* To note a few: Extinction is both a biological norm and a catastrophe that casts biological norms aside. As Darwin showed, there can be both a proliferation of species and an intensification of extinction at the same time. Extinction is central to evolutionary processes but also the radical end and undoing of these processes of speciation. The science of extinction contributed to both the consolidation and the emptying out of the category of species. The demise of a species is a measurable and calculable event, and this empirical data has served a fundamental role for comparative anatomy, paleontology, and population ecology; yet each extinction is an immeasurable and incalculable absence that cannot be encompassed by statistical science. The broad connotations of the concept of ecology have quasi-utopian associations of an assembled commons of life, what Darwin metaphorically called "an entangled bank." Extinction has undeniable dystopian connotations, and its metaphors and landscapes are consistently envisioned as bleak. However, evolution, ecology, and biodiversity are made and unmade in conditions of species precarity and "injury" (Darwin's word) that can never rule out extinction in the midst of such entangled relationships. Such thinking about extinction leads to an expansion of consciousness in the recognition of the end of consciousness.

*Extinction presents both an ecological demand and an ecological dilemma in caring for endangered life.* There is no ecological scenario in which there are no more extinctions at all, yet no extinction today should be deemed permissible in advance, and every impending extinction implores prevention. Ecology is both a descriptive science and an ethical project inclusive of creative efforts that imagine a better world for all life. Engaging with extinction as ecological practice, however, does not mean trying to redeem or save animals from the difficulties of life or death as such. Caring for endangered animals entails working to preserve them from perishing due to anthropogenic causes without insisting on salvational outcomes and without idealizing ecosystems as danger-free utopian spaces. Studying the history of extinction and vigilance toward preventing extinctions in the present does not commit one to the restoration of an ideal garden of protected and enduring life, nor does it necessitate a dystopian vision in which nothing matters because extinction renders everything meaningless in advance.

An ecology of extinction includes urgent care for endangered life and, at the same time, a questioning of how care is arranged, distributed, and integrated into the global biopolitical economy. Ecology is the problem of how species coexist within ecosystems that intermingle shared realities of life and death. An ecological sense of extinction involves examining how life and death continue to be contested and changing terms, rife with biopolitical decisions of who lives and who dies, which species are favored and which are neglected or exploited.

*Understanding extinction requires reactivating existential thinking to comprehend the limits of life.* Existential categories are fundamental categories of subjective, lived experience, spanning the range of birth to death, including flourishing and failure. These lived experiences are core conditions for the existence of any species, not just humans. The existential condition is not the same as the biological domain. In existential thinking, life and death are "conditions" that are cultural as well as natural, experiences as well as factual verities, horizons of understanding and orientations of meaning as well as material realities. Terms such as *lived experience, precarity*, and *finitude* mark the suturing of life as a phenomenon that requires interpretation, choice, and action to life as biological fact. Furthermore, existential philosophy elevates the finitude of thought and life into primary philosophical concerns even while finitude resists and unravels thought and experience. This philosophical method for attending to mortality and precarity that affect all existences must be applied to situations of endangerment and extinction in specific times and places, examining how speciesist metrics and hierarchal norms for human and animal bodies has led to historical instances of life put into peril.

Throughout this book, the label *existential* does not refer to any specific philosophical school or philosopher. However, it is worth noting that existential philosophies emerging in the nineteenth and twentieth centuries arose at a time when philosophers did have knowledge of irreparable species loss. Existentialist philosophers, as witnesses to a number of catastrophic global events destructive of subjectivity and life processes, including world wars, colonialisms, genocides, and pandemics, developed a thinking of care, lived experience, and finitude as core philosophical projects. These thinkers recognized that biological life and existential meaning are without absolute foundations or ontological guarantees. Nothing guarantees the safety, security, and continuation of an individual or species. There is no absolute of Nature, march of Spirit, mandate for Being, or metaphysics of vitalism that maintains and perpetuates meaningful species' lives. Extinction is an existentialism because the end of a species is not just an empirical and biological event but also a re-configuring of the available forms of being and meaning. The loss of a species, as the permanent subtraction of a life form and the transition from being to nothingness, changes the entire scope of the ontological space for possible lifeworlds.

*Thinking about extinction includes cognizing the limit ends of the concepts and forms we use to engage with such loss.* Every account of the end of species is forced to register the inadequacy of any cultural form for such radical erasure. Each account also raises questions about what happens to form itself when representing and engaging with last and lost lifeways. Since extinction is the disappearance of a lifeform, how can any cultural form or medium signify that very disappearance? How can a representation show the radical absence of the representational thing? How does one address the loss of address?

There are many compelling recent studies of the genres used to narrate apocalyptic plots,[62] yet further analysis is needed to comprehend what extinction does to the very forms and devices of narrative. In texts depicting either human or nonhuman animal ends, every sentence is intensified to a distressing degree, as such sentences carry the impossible possibility of last witnessing. Ordinary language and emotional expression are implicitly felt to be not powerful enough to register the magnitude of the disappearance of a species. Even in the early natural scientific discussions of extinction, the language used often carried an extreme sense of upheaval and melancholy (Cuvier's *Theory of the Earth* demonstrates an early example of this search for extreme language to account for extinction, using terms such as: catastrophe, deluge, revolution, convulsion, calamity, and "intestine wars"). Every account of extinction is involved in a hyperbolization and extremification of its terms and tropes. Narratives of total human extinction are a genre predicated on imagining the end of genre. The confrontation of a rhetorical device with its limit ends is characteristic of the entire realm of forms and devices used in extinction narratives—address points to the end of address, the reader faces the absence of all readers, language turns on itself into silence, and so on.

*Every extinction calls for an ethical assessment and response to species loss, but, in an extinction event, the means by which ethical judgments are made face their own crises.* In reckoning with each unique event of extinction, it matters how a species died, what can be known as having helped or hurt the species, how humans were or were not involved, and how one can evaluate proper responses to endangered species prior to their extinction. But extinction events also can disrupt or even shatter ethical practices, norms of care and conservation, and established ways of responding. The loss of an entire species includes the loss of relationships the species had established. The end of a species can evacuate the relationships and values previously sustained by that species. Is there, then, any way to make a rigorous values analysis—including determinations of subjective or anthropocentric values (such as a species' contribution to ecosystem services) or objective, biocentric values (a species' intrinsic "natural historical value"[63])—in the extreme case of the total loss of a species? Should all costs be ignored to prevent the most radical outcome of the absolute disappearance of a species? Which norms for species—evolutionary, economic, political, psychological, social, even aesthetic—should be bypassed or prioritized to avoid such loss?

Assessing ethical responses to extinction calls for a close scrutiny of how the concepts and practices of value and care attached to species are themselves severely challenged or even ruptured in the event of such radical loss. As noted earlier, Balzac, after coming to terms with Cuvier's evidence of so many perished species and lost worlds, felt both invigorated and devastated. Encountering an extinction event through representational media, as in a novel, a film, or a museum, can revivify the viewer to value biodiversity in the present but also can leave a psychic wound that cannot be alleviated by any common coping mechanisms. Ecological ethics circulates around the work of mourning, which involves interiorizing the loss of something we care about while also admitting that grief is unresolvable and that we are undone by the haunting and harrowing effects of the knowledge of species death that includes our own finitude.[64]

## Reading Extinctions

This book examines a series of key revaluations and turning points in the history of thinking and responding to extinction events. The chapters are organized in a chronological order to focus on an event of animal or human extinction that contributed to a redefinition of the meanings of extinction at a specific historical moment. In each chapter, I also discuss how cultural works that document and interpret these extinction events draw special attention to the biological and ethical concerns as well as the formal and conceptual problems and exigencies that arise in situations of the end of species. My methods are interdisciplinary across the sciences and humanities and employ conceptual tools from animal studies and environmental humanities fields. Ecology is itself interdisciplinary and methodologically multitudinous. Extinction events, like any ecological event, are instances that must be understood in terms of both nature and culture or, rather, nature-cultures. Ecological thinking responds to how nature and culture are interconnected and co-define each other, yet also disrupt each other in ways that undermine exclusivist claims made in the name of the natural/unnatural. Furthermore, definitions of species and extinction are not just determined by humans; humans and animals co-interpret and exchange with each other and, thereby, share each other's worlds (though not always in mutually nurturing ways). Accordingly, this book employs methods of interpretation that emphasize the imbrications and entanglements of species with each other and with human frames of knowing and engaging with endangered life.

I have chosen a series of case studies that cohere, not because they are the sole important examples of the redefinition of extinction over the past few centuries but because each event provides a crucial window on an intensely charged scene of last life that resonates across understandings of species finitude. However, this book covers just a few historical instances of extinction and limits itself primarily to well-known species or "charismatic megafauna" (mostly large mammals), which are only a small part of the massive wave of extinction now occurring. My own study risks reconfirming an over-emphasis on animals that humans most care about, a position that can perpetuate a kind of anthropocentric species hierarchy that may lead to further neglect and extinctions. My main rationale for this choice is that the terms of extinction have been constructed largely in connection to these highly visible and dramatic instances of animal and human extinction events (I discuss further in chapter five the limitations of focusing on charismatic species). Historically specific definitions of extinction—and even contemporary conservation efforts—overwhelmingly draw from these kinds of extinction events.

Chapter one examines the near-extinction of the bison in North America as one of the most prominent extinction events to occur right after Darwin's work popularized the scientific verification of species finitude. In 1860, there were an estimated 30 million bison across the North American plains, but by the 1880s, the bison was nearly extinct in the wild, with only a few hundred remaining. The killing of the bison was the first major intentional act of animal extinction in an era aware of the scientific knowledge of species death. What also marks this event as a significant turning point in the conception of extinction is that it was documented, photographed, and discussed in "real time" as the decimation was happening. Archival photographs that depict the killing of bison introduced a new kind of photograph and visual culture, the extinction shot, in which the visual and documentary evidence of the end of an animal species also frames how the human gaze is implicated in the meanings of extinction.

Chapter two examines narratives following the life of Ishi, an Indigenous man living in California who was given the moniker "the last wild Indian in North America" by his biographer Theodora Kroeber.[65] The story of Ishi comes primarily through the account written by Kroeber, who used notes from her husband, Alfred Kroeber, the eminent anthropologist who had studied and cared for Ishi while employing him at the Hearst Museum of Anthropology in San Francisco from 1911 to his death in 1916. Ishi lived through the extinction event of his own people, the Yahi of northeastern California. He was cast as a "last man" figure but also romanticized as an original Californian formed by the sun-swept land. This chapter looks at the "last" narratives and imagery associated with Ishi to reconstruct concepts of extinction and idealization connected to Indigenous peoples that circulated in Ishi's time. I also study how psychoanalytic concepts of trauma, Freudian theories of the origins and ends of humanity, and Indigenous ideas of grief became important to Alfred and Theodora Kroeber's understandings of Ishi. He responded with self-confident patience and friendship to being cast as a first and last human figure—categorizations of indigeneity that Ishi refused—which surprised and haunted Kroeber. Ishi's willingness to share his stories and cultivate new relations recasts the notion of extinction with regard to Indigenous peoples into what Gerald Vizenor calls "survivance."

The third chapter examines the rise of the genre of human-wide extinction narratives in the late nineteenth century, an offshoot from early formations of science fiction. H. G. Wells's *The Time Machine* extrapolated on species' ends in connection with the expansion of a public culture of extinction manifested in exhibits in natural history museums and talk in the popular press about planetary disasters. Wells's novel did not project extinction onto an external other but imagined England, and by extension all of humanity, implicated in a future evolutionary decline. Wells's text develops a plot for extinction after debates on Darwinism, and also shows how extinction affects the formal and structural aspects of novels more broadly. Wells's example demonstrates how "last human" novels engage with the narratological limits of form and character at the same time as they give a fictive account of the end of the world. Using contemporary narrative theory, I discuss how last human novels function paradoxically as cultural works envisioning the end of culture. The genre allows readers to consume their own finitude, as well as the end of literature as such, as an aesthetic event.

Chapter four discusses concepts of human and animal extinction circulating during the Holocaust. The Nazi regime declared that the only way to prevent the extinction of the pure German race was to pursue the extinction of the Jews and other non-pure life. The Nazis placed enormous emphasis on this view of extinction as being a central causal force in everyday national agendas. Already, in 1933, at a time when the German population set a new peak at 65 million inhabitants, the Nazis claimed that German "Volk" life faced imminent extinction if the German nation did not act to correct its own biological downfall. This argument was supported by some of the most knowledgeable biologists in Germany at the time. To stave off this purported extinction, the Nazis asserted a biological politics according to which only a strictly "Germanic nature" could exist in Germany. However, Germany had lost much of its wildlife in the century prior to the time the Nazi Reich took power. I also examine how, at the same time the Nazis pursued extermination for Jews and others deemed biological threats to Aryan life, they sought to repopulate extirpated species and de-extinct animals identified as "German," most famously the aurochs, which went extinct in 1627. In the combination of biological and racial zealotry, messianic fervor, and celebration of ultraviolence, the Third Reich sought to change radically the definition of extinction across the globe.

The fifth chapter shifts to contemporary extinction events currently playing central roles in redefining conceptualizations of species finitude. I discuss the conditions of bison, tigers, and coral as emblematic cases for engaging with extinction today. These animals face a situation in which they are not under an immediate extinction threat, yet their populations remain at historic lows. Picking up the story of bison extinction from chapter one, this chapter opens with an analysis of how the recovery of the bison in recent decades has been trumpeted as a success for conservation. However, the bison have returned primarily as animals amenable to economic gain, as food items, and as figures for nationalist restoration. In the situation of tigers, the iconicity of the animal for conservation cannot be separated from the national status of the wild animal in places such as India, Russia, and China. The wild tiger's charisma is inextricably tied to the "exotic" status of caged and privately-owned tigers in North America, even as these domesticated animals are viewed as not having viable roles in the species' survival plan. The tiger is treated as both a last animal (wild tigers) and a mass animal (domesticated tigers). Corals also are fast becoming a barometer for the precarity of oceanic ecosystems in the Anthropocene. As evidence of increased bleaching and mass coral die-offs have accumulated, coral's life and death has now become a central concern in the sixth mass extinction.

Chapter six provides a critical reflection on recent efforts to develop de-extinction technologies to bring back iconic lost species whose genomes can be potentially reconstructed in laboratories. New biotechnologies and efforts to cryogenically store embryonic cells of endangered animals toward developing de-extinction science are changing the definitions and meanings of extinction. Supporters of these projects employ an assortment of environmental and ethical rationales to support their work. Discussions of the ethics of de-extinction have yet to consider fully how this science would change the ecological and existential structures of life by recasting species precarity as a technological and programming problem. De-extinction, at first applicable to a small handful of recently lost animals, must be critically assessed for its ties to broader ambitions for life enhancement and salvationist solutions to the problem of death itself.

The conclusion examines the conjunction of concepts of extinction and space exploration, focusing on Trevor Paglen's recent "last photography" project, a curated archive of photographs affixed to a satellite orbiting Earth. Several notable space scientists and entrepreneurs have stated that humans must achieve interplanetary settlement as a way to decrease the chances that humans will become extinct should some catastrophe happen to Earth. Paglen's photographic plate on a communications satellite is in dialogue with previous messages that have been attached to spacecraft, most famously the gold discs on *Voyager I* and *Voyager II*. Paglen's archive is unlike those gold discs in that it does not present a benign humanist greeting. Some of the photographs depict violent scenes, some are emblematic of a troubled Earth in Anthropocene times, some show scenes of tenderness and beauty, and some, ambiguously, do not point to any summative stance on humanity. These "last pictures" call on witnesses to reflect on their own last gazes in thinking about extinction.

## Elegies of Representation

Extinction implicates humans and animals in a common condition of precariousness that is foundational to what Hannah Arendt called the necessity to "share the Earth."[66] Arendt insisted that, prior even to the establishment of justice and politics, there is a pre-political condition upon which all politics is based. Prior to any political decision or action, we must accept each other's presence and coexistence as constituting the condition of a fundamental plurality on a singular planet. Since Earth is our only habitation, Arendt argued, we must be committed to share it. In her reflections on the Adolf Eichmann trial, Arendt reasoned that Eichmann deserved the death penalty because he and the Nazis violated this principle in their attempt to decide who gets to reside on the planet. No one should be able to decide for everyone who gets to belong to this plurality: "No man can be sovereign because not one man, but men, inhabit the earth."[67] Arendt's precondition for politics now has become its own political project. It will take an enormous effort to ensure the ongoing sharing of the Earth. Arendt herself did not specifically remark upon the commitment to share the Earth with animals and all other nonhuman life. Furthermore, Arendt rejected any naturalist defenses of human rights or attempts to ground political laws in nature. Yet sharing the Earth is not just a human project, and the human condition also is an ecological condition. Arendt's work can be applied toward the premise that humans should not be the ones to decide what species get to participate in planetary plurality.

As we cultivate the arts of sharing the world, we face the need for the arts to reckon with the loss of world and the diminishment of the planet's plurality of biodiverse life. Each chapter of this book focuses on natural and cultural extinction events that are instances of the unsharing of the Earth. Cultural responses to extinction can, paradoxically, help us share this unsharing. However, cultural works that contend with extinction face questions about their own ethical and interpretative crises. Debates over how to interpret artworks on extinction reveal different ways forms and genres engaging with last animals become implicated in their own ecological and existential dilemmas. W. S. Merwin's poem "For a Coming Extinction" has garnered significant notice and critique from environmental humanities scholars for its relatively early poetical attention to the devastation of animal life. Taking this poem as an example, however, shows how critical tendencies to assess art works in terms of what makes for an appropriate reaction to extinction do not fully grapple with how representing extinction troubles both the aesthetics of form and the judgments on the ethical effectiveness of these forms.

Merwin's poem positions a human speaker addressing the gray whale as the species edges closer to extinction to relay a bitter message:

> Gray whale
> Now that we are sending you to The End
> That great god
> Tell him
> That we who follow you invented forgiveness
> And forgive nothing[68]

The opening stanza, a direct address to the last of the gray whales, sounds as if the poem will blithely dismiss the cetacean, which was commercially hunted in the nineteenth century and saved from the brink of extinction by international protections established in the mid-twentieth century. The gray whale has since made a dramatic comeback and just recently was categorized as "least concern" by the IUCN. But this recovery is not imagined in the poem, which offers a parting message to the gray whale that is brusque and saturated with human self-importance. The second and third stanzas seem to temper this attitude by drawing from different reactions to death, with a mix of dismay, empathy, and a reference to "bewilderment." The poem connects the decimated whale species to the ghosts of others, "The sea cows the Great Auks the gorillas," described chillingly as "sacrifices" (68). The final stanza of the poem reverts to the initial frame of human arrogance as the speaker enjoins the whale to tell "that it is we who are important" (69). There is no punctuation throughout (there is almost no punctuation in the entire book), which amplifies the syntactical role of line and stanza break, suggesting a disorienting kind of visionary poetics that is unable or unwilling to summarize itself with syntactical norms.

In her reading of this poem, Ursula Heise charges Merwin with "cut[ting] off any possibility of response"[69] for the gray whales and for any favorable human interaction with nonhuman life. "The whale's death turns out not to mean anything other than humans' grim and deadly self-affirmation" (47). She finds the poem to be caught in an unconvincing elegiac mode stuck in foreseeing only what Heise calls an "environmentalist rhetoric of decline" (8). As Heise points out, the rhetoric of a cursed inevitability and diminishment embedded in many elegiac accounts of extinction can perpetuate the view that biodiversity loss is an implacable tragedy no matter what humans do. She adds, "Important as the genre of the species elegy has been for mobilizing public support, conservationists have also had to face its limits. The elegiac mode tends to leave out species that cannot easily be associated with particular cultural histories, and its nostalgic and pessimistic tone puts off many potential supporters" (50). Instead, Heise makes the case that a wider variety of narratives and less pessimistic cultural productions might spur more attention to imagining new and more effective ways to respond to biodiversity loss and foster further care for endangered species.

Heise's critical assessment befits the attitude of the poem's speaker, yet it is hard to imagine readers finding the voice of this poem as straightforwardly their own. The condescension of the poem with its direct address issuing imperatives to the gray whale is palpable—"tell" that "great god" who is not so great after all to have granted humans the power of forgiveness as well the capacity to disavow it. The speaker's self-important voice rankles the reader who, like the gray whale, is being told what to say and how to feel about it. Merwin wrote the poem during the escalation of the Vietnam War, and included it in his volume *The Lice* (first published in 1967), which featured a number of poems excoriating the rampant and shameless acts of violence and desecration unleashed by humans against each other and against nonhuman animals. Hank Lazer calls this book "an elegy for the dying planet" and an account of "uncreation" in emptying the world of life.[70] The same voice in "For a Coming Extinction" that is callous to the gray whale includes itself—"we who follow you"—in its vision of extinction. It is the anti-elegiac blitheness of the speaker that contributes to putting humans, whales, and tropes of visionary poetics on the same path to existential destruction. Paul Sheehan comments that this poem is "more concerned with self-erasure or self-extinction than (as Heise claims) with 'self-affirmation.'"[71] Merwin's poem supplies a series of self-incriminating judgmental statements that divulge their own complicity in the bleak and heedless finality of "The End / That great god." As the speaker's voice implicates itself in its own "coming extinction," the reader must recoil and refuse this fatalistic voice even while recognizing oneself implicated in it.

Poems attentive to extinction often employ tropes that engage with the limit-extremes of tropes themselves. Contemporary poet Eleni Sikelianos's series "How to Assemble the Animal Globe" elegizes several extinct species while also registering the effects of what extinction does to poetic form. The series comprises several poems written for extinct animals across different regions and continents. In the section Islands, there is a poem "Aldabra Brush Warbler," the name of a slender and timid bird from atolls near the Seychelles, an island country off the east coast of Africa. The poem, little more than a series of dates, records the first sighting of the bird and other important observational dates up to its extinction, most likely because of introduced predators, including cats and rats.

"discovered" in 1967
  described in 1968
  lost in 1969
  found in 1975
  gone in 1983[72]

The sightings of the bird signify as ornithological data and also inscriptions for an epitaph. The bird makes no direct appearance; it manifests only as the fleeting shadow of its infrequent observers. The poem presents only a minimal voice and address, demurring from commentary and elegiac adornment. Sikelianos's poem does not strive to perform the lastness of voice; rather, it delivers a minimalness of address and isolated dates of furtive contact with the bird until its extinction. Instead of a visionary take on extinction using an apocalyptic or prophetic address, this poem conveys no confidence in the elaborations of elegy or in poetic judgment to know what to say at such scenes of radical loss. The paucity of reference offers little more than a bare ruined choir where late the bird sang. The poem does not tell the reader what to feel but knows that the scene of the end of a species will be charged with the most intense affects and extreme forms of address. The flatness of the poem acts as a mirror and a stone, but not a guide, on which the reader cannot help but project a range of emotions that extinction elicits.

Genres of elegy, apocalypse, and tragedy are commonly used in representations of extinction, and these genres support multiple, complex, and often contradictory responses and readings. Apocalyptic genres, even as they provide an immediate aesthetic shock, also can help make traumatic visions of finitude more legible and manageable. The poetic elegy, as it bids death and finitude to enter into art, is not a static genre but, like all other genres, is always "evolving, hybridizing, self-subverting."[73] Another way of phrasing this is that one need not always read elegies strictly elegiacally. One role of elegy is to make death consolable and meaningful by situating an individual death in the context of an ongoing social and natural communities. For much of the history of elegy, it was conventional to situate the scene of mourning within a pastoral setting to connect cycles of human life and death to nature's seasonal cycles of spring profusion and winter torpor that will carry on in future generations. Extinction events, however, disrupt and dissolve generational cycles. The elegies of extinction today, in times of global heating and anthropogenic climate distress, can rely no longer on the traditional cyclical consolations of the pastoral. We still use the elegiac mode to understand extinction, but such radical erasures of life in times of mass species loss also change the formal and environmental conditions of elegy.

However, becoming knowledgeable about extinction does not commit us to a single track of doom and despair, even in ecologically distraught times. Elegies, as expressions of loss and love, can both stun us and transform us. Chickasaw writer Linda Hogan movingly speaks of the need to look back and look ahead to account for the emptying of life and determine a way forward: "There are already so many holes in the universe that will never again be filled, and each of them forces us to question why we permitted such loss, such tearing away at the fabric of life, and how we will live with our planet in the future."[74] Responding to these questions, art works can help to mediate between the deep time and the present time of species, and between changing conceptions of extinction and singular encounters with endangered life.

This book provides an examination of several instances of the redefinition of extinction by bringing together multispecies studies with media-specific formalist criticism to comprehend what happens at the end of a unique form of life. As extinction rates rise, we are witnessing a great unworlding and de-pluralization of the Earth, a planet becoming more and more finite. We are witnesses to a great wave of life and its crash. Heeding what Deborah Bird Rose calls the "ecological existentialism"[75] of extinction presages neither a definitive dystopian nor a utopian horizon for all species, but it does commit us to becoming more self-aware of how being on Earth entails the fundamental condition of being-together with other life forms—the last best hope for learning how to share the planet.

PART I

# 1 /    Photographing the Last Animal

The emergence of a science of extinction in the nineteenth century occurred at the same time as the emergence of new documentary technologies, including photography. Ever since Darwin explained how the
permanent loss of species played a central role in natural history, the techniques and media used to document extinction have taken on prominent roles in the scientific and public knowledge of last animals. Interest
in photographing rare and endangered animals contributed to the rise,
in many different local and national sites across the globe, of a public culture of extinction at the end of the nineteenth century. By the late
nineteenth century, the ends of species became the subject of everyday
conversations and viewable in images circulated in a variety of print media. Natural history museums organized exhibits around disappeared
animals, while a select number of photographers turned their lenses to
taking images of those wild animals whose populations were collapsing.

This chapter examines how photography made by white settlers of the
dwindling wild bison across the North American plains during the late
nineteenth century became imbricated in contested responses toward animal extinction, conservationist attitudes, and settler colonial biopolitics. Photographers began to bring their cameras into the regional habitats
of the North American plains for the purpose of photographing declining
wildlife and the hunters involved in their immediate demise, furnishing
visual representations of Frederick Jackson Turner's declaration that "the
frontier has gone."[1] These photographers consciously sought to capture
this "gone-ness" with explicit references to the "last" bison hunt. To do

so, they had to make visual sense of the scene of extinction to decide how to photograph the last of a species.

As early as 1843, John J. Audubon warned: "Like the Great Auk, before many years the Buffalo will vanish. Surely this should not be permitted?"[2] Harvard University zoologist Joel Allen, in an extensive 1877 study of the bison (also called buffalo[3]), declared "the present decrease of the buffalo is extremely rapid, and indicate most clearly that the period of extinction will soon be reached."[4] In this chapter, I discuss a selection of bison hunt photographs that mark one of the first moments the camera played an important role in how an extinction event became recorded and disseminated and stoked public concern. I describe these photographs of last animals as "extinction shots" and examine how these photographs engage with the animal's disappearance in the context of settler colonial ambitions to dominate Indigenous peoples and control and monetize the plains ecosystem. I also analyze how these images contribute to the theoretical discussion of the photographic medium itself as fundamentally engaged with death and loss at the ontological and formalist level of the photographic image.

Since Darwin, the history of conceptualizing and documenting extinction has been closely connected to the capturing and viewing of images of last animals. Photographs of lost species inform definitions and reactions to extinction, foremost in the way these images raise the question, to borrow a phrasing from John Berger, Why look at extinct animals? Berger remarks that one must look at animals because "everywhere animals disappear" in an "enforced marginalization."[5] In the late nineteenth century, the camera helped make uniquely visible and legible what it means for a species to reach the extreme condition of finitude and final invisibility, while also framing complicitous human involvements in an extinction event such as the decimation of the bison.

An estimated 30 million bison roamed the plains of North American until the 1850s.[6] By the mid-1880s, fewer than 1,000 plains bison existed. The devastation of the plains bison due to widespread hunting, primarily during the 1870s, was one of the first extinction events to be made visually accessible by photography. Although, in this case, a total extinction of the animal did not occur, the decimation of the bison provoked one of the first widespread discussions about the meanings of extinction on a global scale. Many factors make this extinction event particularly noteworthy, including the large amount of visual culture made during the decades of bison devastation that became part of the conversation around what was happening to the animal. Photographs of the bison taken in

North America at the end of the nineteenth century depict the eradication of the animal as the hunt was happening. These photographs are "about" extinction, in that they document the apparent last days of the plains bison. But they also are about extinction because they compel the viewer to use the visual elements of the photograph to make sense of the loss of the animal, the processes that contributed to the animal's demise, the complicity of such images with power over Indigenous lives and land, and the role of photography and its image technology in witnessing the last of a species. These photographs also inform us about how the history of conceptualizing extinction intertwines with the representational paradox of how to make the disappearance of the animal visible.

The imagery of the plains bison produced by nineteenth-century settler artists prior to the arrival of photographers in the North American Midwest typically shows the animal's incredible abundance and its dashing pursuit by Native American or white settler frontier hunters. Perhaps the most common visual trope of the bison in mid-1800s paintings and drawings is of the herd and the stampede, the huge, rolling, moving mass of bison thundering across the plains. In William J. Hays's "The Herd on the Move," there are no other visual cues on the plains other than the apparently endless bison stretching to the horizon (see figure 1-1). Here, the bison, which seem to emerge literally out of the land itself, are

FIGURE 1-1. William J. Hays, "The Herd on the Move," colored lithograph, 1862, Mabel Brady Garvan Collection, Yale University Art Gallery.

converging and heading straight at the viewer, who is about to be trampled. The viewer feels thrown into the action, and with no humans in the painting serving as mediators, the spectator stands as if in direct witness to the tremendous power of the endless herd.

George Catlin's "Catlin and His Indian Guide Approaching Buffalo under White Wolf Skin" (1846–48) is one of a number of similar images Catlin painted while traveling across the plains of the United States. In this painting, again, there is nothing on the plains, not even a tree, save for the long stretch of the bison herd that fades into the background. The sparseness of the hills makes for a stark contrast to the abundance of the bison. The simple trick of wearing wolf skins allows hunter and painter to approach the bison to kill or to draw them at close range. Catlin's unknown Native American guide carries a bow and arrow, a throwback to an antiquated means of hunting then. Catlin appears as the "contemporary" figure with both a gun by his side and a sketch book in hand. There is no other context provided for this hunt/drawing session, but the image suggests the unsuspecting bison can be "rendered" by artist or hunter at will. Such scenes are evocative of what Antoine Traisnel identifies as an expansion of a system of "capture" as a new animal condition, one that follows from a combined aesthetic and economic power over animals "inextricable from the making of the new nation—the construction of a hegemonic American identity and iconography and the consolidation of early capitalism and settler colonialism."[7]

Hays's and Catlin's paintings, along with many other works of similar subject matter, became invitations to romance the frontier as ever open, teeming with the massive mammal, beckoning the hardy hunter and homesteader. The bison—the largest living land animal on the continent (males average 2,000 pounds)—appears to be limitless in number and unlimited in its exposure to the frontiersman's gaze. Yet by fusing the gaze of the hunter with the artist, these images already indicate a blurring of life and death for the animal.

The political, economic, and ecological factors that caused the near-extinction of the bison have been well documented by historians and naturalists.[8] The numbers of bison roaming the North American plains in the 1840s remains hard to determine, but most researchers agree that a population of approximately 30 million could be supported by the food available in its remaining native range that stretched from Texas up to the northern reaches of Alberta, between the Rockies and the Mississippi River. By that decade, the number of Indigenous communities subsisting on the bison for food and clothing had increased due to forced migration to the plains.

Still, there was little noticeable change in the overall ecology of the bison until after the American Civil War, when the presence of frontier hunters increased and cattle ranchers began moving in, spreading disease among animals and increasing competition for grasses.[9] The key technological developments of more accurate and quick-firing rifles, along with the construction of the transcontinental railroad in the United States beginning in the early 1870s, made it possible for hunting for bison robes to occur at a mass industrial scale. Between 1872 and 1874, the total shipment of buffalo robes to the East numbered approximately 1.4 million (this number includes only the usable hides; it is estimated that skinners managed to obtain a usable hide in only one of five kills).[10] In Dodge City, Kansas, initially the capital of the hide trade, the largest hide dealer in 1872–73 shipped over 200,000 hides in one year.[11]

A good hunter with a small team of dressers could kill fifty bison a day (some hunters claimed to be able to shoot well over a hundred on some days). During the herding season in late summer, the only time bison would congregate in mass, in a period of three months, a hunting team could amass 2,250 hides.[12] There were several thousand settler hunters in each state, plus Native Americans who hunted for their own livelihood or to sell robes on the market. The bison slaughter was astounding in terms of scale, speed, and technology, and executed by a relatively small number of participants spread in small hunting camps tied into a global supply chain. New tanning techniques were invented by the mid-1870s that made bison hides, previously stiff and heavy, into more supple and versatile leathers. In addition to supplying material for clothing and furniture, one of the main uses of this leather was for belting that was highly desired by industrialists for increasingly complex wheels and roller systems in the burgeoning factory assembly line.

The slaughter of the bison happened too quickly to bother gathering the meat of the animal in most cases (usually only tongues were culled). The meat was left on the corpse and allowed predatory animals to flourish, particularly wolves, which, in turn, led to further bison depletion (a bull carcass, on average, yielded 550 pounds of meat). In the southern Midwest states, the collapse in bison numbers was apparent by 1875. In the northern states and Canada, hunters harvested several hundred thousand robes consistently in the 1870s until 1882–83, when the numbers of bison hides shipped by train dropped significantly. By 1884, the hunt was over. Fewer than 1,000 bison remained, including a small herd in Yellowstone National Park, which forbade hunting, and small scattered groups on a few isolated ranches and zoos that sought to domesticate the animal.

## Picturing the Bison Hunt

When photography of the North American plains began to become a genre of its own after the American Civil War, the photograph of the bison was constrained to depict a very different figure. Photographs in the nineteenth century of bison herds on the move are nonexistent, due to a variety of factors, including the rarity of photographers living on the plains, the preference for static shots of landscapes and portraits, and the fact that the wet collodion process commonly used until the early 1880s required a small and fragile lab of chemicals to be carried with the camera, hampering the photographer's mobility.[13] All these factors reduced the likelihood that a photographer would encounter a bison stampede, let alone with a camera that was technologically sophisticated enough to take a picture of a moving animal.

Instead, bison photography in the nineteenth century overwhelmingly shows the animal already killed and framed within the context of the rhetoric and iconography of "last" hunts. The recurrent visual trope of the static bison, whether being hunted, skinned, dried, or taxidermied, orients the viewer of these images toward the inevitability of the animal's killing. Some of these images stress the subjugation and overpowering of the massive animal before human onlookers. Others imply in their portrayal of placid, quiet scenes with an air of the pastoral—in which a killed animal lies tenderly at the foot of the image, as in F. Jay Hayes's "Buffalo Hunting, Montana" that the animal appeared to go quietly and conveniently to its destruction. In this photograph (see figure 1-2), two hunters sit on horseback in a relaxed fashion behind the bison. The hunt is not being romanticized in this image; the hunters mean business. The close-cropped perspective of the hunters and kill provides no sense of grandeur or tragedy in this matter-of-fact image of the end of bison hunting.

In all these images, the killed animal is tipped on its side or back—without its footing, the animal shown in every one of these photographs is already defeated. The technical reasons behind photographing bison in stilled positions are converted ideologically by these photographers into imaging these animals as isolated and spiritually subdued, shorn of their power and social and ecological connections.

Another common visualization of the bison observable in photographs taken in the 1870s and 1880s concerns the animal's abundance as commodity form, its skin and bones heaped for monetary gain. In a photograph dated August 9, 1890, and credited to H. C. E. Lumsden (see figure 1-3), a man stands next to a wall of bones in Saskatoon. The bones

FIGURE 1-2. F. Jay Hayes, "Buffalo Hunting, Montana," photograph, 1882, Montana Historical Society.

FIGURE 1-3. H. C. E. Lumsden, "Piles of Buffalo Skulls at the Railway Siding, Saskatoon, Saskatchewan," photograph, 1890, Glenbow Museum.

are stacked neatly, awaiting shipping at a train depot. The wall of skulls is angled in a foreshortened perspective, stretching to meet the horizon line. The human figure here provides a sense of scale, holding a long tree branch to show the height of the bone wall. The man appears calm and well dressed. He is positioned directly under a section of the wall that has horns protruding into the sky, which, perhaps unintentionally, adds a demonic tinge to the scene.

Echoing in inverted form the tropes of animal plenitude in paintings from earlier in the century, these photographs of mass collections of bison remains show the animal as an industrialized object. The visual culture of the mass-processing of the animal body as lifeless, stackable goods evokes in contrastive form the paintings of bison herds flowing across the hills. The visual power of thundering animals is mirrored in negative in photographs of the soundless presence of an empire of bones. What we see with these photographs of the small- and large-scale operations involved in the killing of the bison is the meeting of images of the vanishing of the bison and the development of the industrial animal-rendering economy on the plains. Both types of bison photograph reveal scenes of the organized division of labor that will feed large-scale economies of animal consumer goods.

What does it mean, then, to watch extinction as it happens, and to document that watching? How does one photograph the disappearance of an animal, the absence of the visual referent? The development and spread of photographic imagery of bison slaughter is exemplary of an early historical moment of public discourse becoming tied to self-consciousness about animal extinction. Since the demise of the bison, the photography of last animals has played a key role in the development of cultural attitudes around animal finitude.

The visual record of the destruction of the bison contributed to the distinctly recognizable trope in photography that can be called "the extinction shot," defined by the many ways of attempting to visually capture the last animal of a species. The extinction shot is not a singular visual perspective, and the photographer need not know in advance that an animal photograph may end up being part of the historical record of a last animal. There are multiple kinds of extinction shots that can contribute to ways of knowing and coming to terms with an extinction event. Extinction shots can image the last days of a particular animal, but they do more than that, as they provide a framing for confronting broader notions of lastness and species finitude. These images allow for the undoing of species to be seen in a single glance; they also invite and call the viewer to reflect on the desire for "last chances to see." By incorporating a sense of lastness into the image, extinction shots also raise reflection on the limit ends of image-making. In this chapter, I discuss, in particular, some photographs taken of the final years of the bison hunt, when hunters knew the animal was nearly wiped out. Since the hunters and photographers understood they were chasing some of the last bison, the photographs of these last animals were taken with an eye toward the finality of extinction.

When Darwin wrote *On the Origin of Species* in the 1850s, the science of extinction already had achieved some stature and consensus regarding the mechanisms and permanence of extinction. What Darwin added was an emphasis on theories developed out of real-time observation of animal behavior amid "many complex contingencies, on which the existence of each species depends."[14] Darwin's modeling provided the basis for being able to watch and document extinction as it was occurring. He saw human induced extinction events as no different in kind than other forms of extinction; indeed, anthropogenically-caused extinction provided direct evidence used to model a more general understanding of the processes of a species' demise. Reviewing the geological record of fossil discoveries, Darwin wrote: "We see in many cases in the more recent tertiary formations, that rarity precedes extinction; and we know that this has been the progress of events with those animals which have been exterminated, either locally or wholly, through man's agency" (258).

Darwin's "we see" and "we know" emphasize that rarity and extinction, "either locally or wholly," can now be watched, recorded, and understood as it happens due to "man's agency." Following on Darwin's observations of the courses of extinction in the present tense, the decimation of the bison marks the first time in history that animal extinction was witnessed and debated in real time at a national and international scale. Hunters, frontiersmen, politicians, animal cruelty patrons, conservationists, and Indigenous peoples on the plains were conscious of being spectators and participants in an extinction event as it was happening.

Although Darwin points to a few specific cases of extinction in natural history—ammonites, for example—he provides no extended documentary or narrative account of the death of any one species in depth. Yet in Darwin's lifetime, a new trope had emerged—what can be called the trope of the last animal—that had become increasingly nuanced by the time he wrote *On the Origin of Species*. Such tropes permeate cultural works dedicated to telling fictional or nonfictional stories of the decline and end of a species (early examples include Hugh Strickland's nonfictional *The Dodo and Its Kindred*, and, in fiction, Cousin de Grainville's *The Last Man* and Mary Shelley's *The Last Man*).

Instead of treating the dwindling few of a species as perfunctory, the last animal trope that burgeoned in the nineteenth century began to put enormous attention and stress on what happens to an animal approaching extinction. With this trope, the terminality of a species became increasingly an identifiable, documentable, and consumable phenomenon. Last animal tropes made extinction legible and thinkable—we will see iterations of this trope throughout the chapters of this book—yet these tropes must be critically analyzed by situating them in political, ecological, and representational histories. The development and adaptation of this trope is part of the process of becoming self-aware of extinction, and it also has contributed to defining the perceptions, affectations, and public knowledge of biological finitude.

In the case of the decimation of the bison, the circulation of last animal tropes becomes emblematic in the photography of the animal made by settler frontiersmen. These tropes of "last" visibility for the bison became embedded in long-standing histories of violence and colonialism toward Indigenous peoples dependent upon the animal. References to the bison's lastness and gone-ness in the latter half of the nineteenth century circulated in the context of declarations issued by the U.S. government intent on confining and pacifying Native American peoples living across the plains. The expansion of the hide trade in North America at industrial capitalist scales converged with the political practice of removing and sequestering Native Americans, who depended upon the animal for daily sustenance.

Recalling Indigenous life ways before this extinction event, Luther Standing Bear detailed the comprehensive nourishment from Lakota bison hunting practices by remarking: "When our people killed a buffalo, all of the animal was utilized in some manner; nothing was wasted. The skins were used as covers for our beds; the horns for cups and spoons, and, if any of the horns were left, they were used in our games."[15] The Sioux doctor and author Charles A. Eastman (Ohiyesa) relates that he was born in a buffalo-hide teepee in northern Minnesota. As an adult, he witnessed how the U.S. government conceived it "impossible to conquer the Plains Indians without destroying the buffalo, their main subsistence. Therefore, vast herds were ruthlessly destroyed by the United States army, and by 1880, they were practically extinct."[16] In an interview, Old Lady Horse (Spear Woman), a Kiowa Elder, recalled the combination of sustenance, ritual, and religious roles of bison: "Everything the Kiowas had came from the buffalo. Their tipis were made of buffalo hides, so were their clothes and moccasins. They ate buffalo meat. Their containers were made of hide, or of bladders or stomachs. The buffalo were the life of the Kiowas. Most of all, the buffalo was part of the Kiowa religion."[17]

The bison furnished both physical and spiritual nourishment, which is why the Lakota leader Sitting Bull stated; "A cold wind blew across the prairie when the last buffalo fell—a death wind for my people."[18] In a recent commentary on Indigenous-led efforts to return bison populations to the plains, Tasha Hubbard, a First Nations/Cree scholar and filmmaker, details how the killing of the bison constituted a genocidal act since the Indigenous inhabitants did not just depend on the animal for food but also forged a kinship relationship with the bison. Hubbard recounts how "Indigenous peoples saw the buffalo as their protector, who took a position on the front line in the genocidal war against Indigenous peoples."[19] Eradicating the bison destroyed the Indigenous communities, ecosystems, and kinship relationships that sustained each other.

The photographs of the demise of the bison, among the first photographic images that attempted to make animal extinction visible, thus, also doubled as images of the "clearing of the plains" of Indigenous lives and lifeways. The doctrine of *terra nullius* used to legitimize settler theft of Indigenous lands is linked to a view of *animalis nullius*, which treated animals as removable, capturable, and eradicable without limits. *Terra nullius*, first established in Roman law and extended through policies of imperialism and colonization, indicated that land not declared as owned and cultivated could be taken possession of and developed. Animals on such lands were "fair game" and permitted to be hunted without restraint. Treating the bison as *animalis nullius* meant settlers could kill and harvest the animal knowingly without any restrictions, even to the point of the annihilation of the species.

By the early 1870s, U.S. government officials were well aware that numbers of the bison were diminishing, and made political calculations based on the knowledge that extinction was likely to occur soon. Columbus Delano, secretary of the interior under President Ulysses S. Grant, declared, "I would not seriously regret the total disappearance of the buffalo from our western prairies, in its effect upon the Indians. I would regard it rather as a means of hastening their sense of dependence upon the products of the soil and their own labors."[20] To this end, the U.S. government provided free ammunition to buffalo hunters, while Grant pocket-vetoed a bill passed by Congress to extend federal protection to the animal by making it illegal for anyone other than a Native American to kill a female buffalo.

Many political and military leaders perceived the extinction of the bison as the latest salvo of an epochal racial war between settler whites and Indigenous peoples. In 1866, after attacks on government forts in Montana, Generals William Sherman and Ulysses Grant urged Congress "to provide means and troops to carry on formidable hostilities against the Indians, until all the Indians or all the whites on the great plains, and between the settlements on the Missouri and the Pacific, are exterminated."[21] The clearing of the plains paved the way for settler homestead farming and ranching that would take over the massive ecosystem and quickly turn it into ranch land and wheatfields feeding the growing demand for meat and grain.

The national drive initiated in the late 1880s to save and protect the bison did not follow from a willful rejection of the colonialist attitude that coupled the clearing of the animal with the domination of Indigenous peoples. William Hornaday, who helmed the efforts to preserve the remaining bison and established a small preserve for the animal in the National Zoo in Washington, DC, espoused bison conservation with the assumption that Indigenous peoples on the plains would have no choice but to accept "the inevitable results of the advance of civilization."[22] Hornaday's efforts at conservation drew from his view that the bison could be of significant economic value to Indigenous and settler hunters alike, if kept in small reserves and controlled areas under a plan of stabilized management. However, Hornaday expressed pessimism the bison could ever survive again on its own in the wild, noting: "The nearer the species approaches to complete extermination, the more eagerly are the wretched fugitives pursued to the death whenever found. Western hunters are striving for the honor (?) of killing the last buffalo."[23]

## Last Bison Photographs

The photographic record of the demise of the bison follows from the eyes of settler hunters and frontiersmen who romanticized the chase while profiting from the increasingly scarce animal. The photographers who followed the final days of the hunt, and who sometimes participated in the hunt themselves, were aware they were photographing an extinction event. The photographs of L. A. Huffman (1854–1931), whose studio was based in Miles City, Montana, provide some of the last images documenting the hunt for the few remaining bison herds in the United States. Huffman was himself an avid hunter who once photographed a trophy shot of himself behind a bison he killed in 1877. Huffman well understood that he found himself positioned at a turning point in the development of the plains and sought to photograph the supposed last days of the frontier and the first decades of the cattle ranching industry and homesteading. His photographs of the Montana landscape, along with images of Yellowstone National Park and portraits of local Native Americans, sold well and influenced other regional photographers and painters.[24]

Huffman's photograph "Killing of Cow and Spikes, Buffaloes in Smokey Butte Country, Montana" (see figure 1-4) depicts a snowy landscape that has become the site of several bison kills. The white of the sky and the snow-swept plains sets the background for the brown-black bison bodies

FIGURE 1-4. L. A. Huffman, "Killing of Cow and Spikes, Buffaloes in Smokey Butte County, Montana," photograph, 1882, Montana Historical Society.

to be seen in full contrast. A rifle is positioned on one of the dead animals in the center of the photograph. By including arms and gear in the frame, the image points to a conflation of battlefield imagery with hunting imagery. One bison has been skinned, with the pelt laid out behind the animal. There are no humans present. The scene has a quiet intensity, showing both death and work. Huffman positions the camera in a way that emphasizes the large head of a killed bison in the bottom of the foreground. The eye of this bison is wide open; the viewer of the photograph is left to contemplate the somber and sober reality of these animals' death. This image confirms that the bison's extinction was not due to Darwinian processes of natural selection but, rather, the result of settler interests applying new technologies in subduing the plans.

Huffman also took a photograph of this same scene before the skinning started. He titled this photograph "5 Minutes Work. Buffalo Cows." This image includes the horse used by the hunter but, again, no humans.

The title casts a pointed judgment at the speed of the hunt, if not the death itself, in wiping out the few remaining small herds.

Huffman's photograph "Taking the Monster's Robe" shows another winter hunting scene (see figure 1-5), this time picturing two hunters in the task of skinning a very large bison. The two hunters flank each side of the animal, providing a sense of the size of the "monster" bison. The caption seems to resonate with heroic deeds, such as Theseus slaying the Minotaur.[25] The bison has been flipped on its back, immobilized in a wholly dominated position with its legs in the air. The caption designates the skin of the bison as a luxurious and fashionable "robe." Only the hide will be collected from this animal; the meat will be left to rot. Huffman's image has the traits of a trophy shot, and there are a number of reasons why these photographs also can be described as extinction shots. Huffman's photos always show the bison already dead, and even if Huffman could take images only of still animals for technical reasons, the fo-

FIGURE 1-5. L. A. Huffman, "Taking the Monster's Robe," photograph, 1882, Montana Historical Society.

cus on bodies after the hunt calls on the viewer to concentrate on the act of killing. These photographs emphasize the lonely death of the bison situated in the sparse, seemingly empty expanses of Montana. It is difficult to gauge the intended affective response to the photograph—the placid bison bodies might evoke sympathy for the animal's demise or might be viewed as a sign of success for the hunters. Some of Huffman's photographs depict the dignity of the bison even after death, while others seem to present the animal as just a large carcass. The images make prominent use of the white starkness of the landscape as backdrop in contrast to the close-ups of the killed bison, forcing the viewer to combine aestheticized visions of the landscape with the sense of brutality and mundane labor involved in shooting and skinning the bison.

The popular magazine *Harper's Weekly* published several drawings in the 1870s that provided a visualization of the extinction of the bison as it was happening. Thomas Nast, widely known for his political cartoons, provided a drawing for *Harper's* captioned "The Last Buffalo" (see figure 1-6). Here, we see a lone bison giving up his robe as a kind of

FIGURE 1-6. Thomas Nast, "The Last Buffalo," ink drawing, 1874, *Harper's Weekly*.

appeasement and surrender to the hunter. The bison hopes to keep its head and its body by gifting its hide to the relentless hunter, who stands shocked but holds on to his gun. This act of submission and acceptance of human domination over the animal is one version of the last animal trope, emphasizing the exhaustion and humiliation of the animal's last days. The bison has no more fight left and strips in order to save itself. The gender of the bison is unclear (both male and female bison have horns), but the disrobing gesture connects to a history of sexual dominance and what Carol J. Adams discerns as the sexualization of meat that subtends patriarchal power over animal life.[26] In the caption, the bison speaks in exaggerated polite overtures: "Don't shoot, my good fellow!" Yet another caption informs the reader "his plea was wasted." The bison's pelt is so deeply commodified that the bison sees no other possible future for existence except to disrobe itself in a violent and naked self-exposure. This image of the last animal exemplifies what Giorgio Agamben calls "bare life," a figure stripped of all social being and reduced to the thinnest of biological life, mere existence.

Agamben applies his term "bare life" to human lives that are no longer granted political standing and are treated as mere corporeal substance whose body is cast into a liminal condition between life and death.[27] Bare life is expelled from both nature and culture, banished into a zone of indistinction that is both within and without the law (politically determined to be exempt from political classification). The disrobing "last buffalo" is forgoing its own state of nature, but the animal's appeal to cultural and political gestures of submission and appeasement are not convincing the hunter.

Agamben states that policies productive of bare life can potentially arise in any political system or event in which persons have "to pay for their participation in political life with an unconditional subjection to a power of death" (90). The production of bare life and the figure of the last animal in much of the nineteenth-century imagery of endangered life starkly resonate with each other, as the last animal is the culmination of treating the species as expendable up until only a few individuals remain, at which point the animal suddenly takes on enormous significance and confronts the viewer from the position of bare or final life. The last animal is revealed as the bare life of the species, as is dramatized in Nast's "last buffalo," a figure of horror and capitulation, confronting both hunter and viewer of *Harper's* pages with its merest of life, its naked flesh shorn of its skin. Nast's drawing evokes how the bison became dispossessed of its body and devitalized in the process of extinction. Instead of furnishing a romanticized vision of vanishing life on the plains, a common theme in the visual culture of the American frontier, this last buffalo is physically and psychologically bereft at its species limits, demoralized and desperate.

## Extinction in Camera

There exists a pair of photographs taken in the late nineteenth century that have become frequently reproduced for their visual commentary on the overwhelming amount of killing involved in the extinction event of the bison. These photographs, made by an unknown photographer, capture what appears to be a typical day at an animal rendering plant, the Michigan Carbon Works in Detroit (see figures 1-7 and 1-8). The images probably were taken some time between 1892 and 1895, well after the end of the bison hunt on the plains. These photographs of a towering wall of bones give an indication of the magnitude and sweep of the hunt. Since the photographs were taken well after the extinction event, the bones are a kind of unintentional *lieu de mémoire* for the decimated animal—a site for memory but not a dedicated memorial. At the same time, these photographs disturb any coherence to the memory of bison life on the plains. The immediate referent of the photographs is the continued capitalization on the animal's death due to the trade in bison bones. The images depict the industrial process of what Nicole Shukin discusses as the material and symbolic rendering of animal into capital.[28]

The bones were shipped by the ton to the Michigan Carbon Works plant for processing, to be rendered into bone-black, which involved baking the bones into a charcoal to be used for filtering sucrose syrups made by sugar refiners, and bone-char, a phosphate- and calcium-rich fertilizer for improving soil. The Michigan Carbon Works expanded rapidly in the late 1880s and early 1890s to become one of the largest industrial plants in Detroit. By 1892, the plant covered 100 acres of land, employing 750 persons, producing over 20,000 tons of fertilizer annually. Neighbors called the site "Boneville."[29]

Looking at the Michigan Carbon Works photographs, the viewer searches for informational cues to interpret this forbidding landscape. On first glance, one is immediately struck by the remarkable orderliness of the pile and the way it takes up nearly the whole visual space of the photograph. The skulls have been carefully and skillfully arranged to form a wall that slopes like the base of a pyramid into the sky. In the image of figure 1-8, the positioning of two human figures gives us a sense of scale.

FIGURE 1-7. Michigan Carbon Works factory grounds in Detroit, Michigan, photograph, circa 1895, Detroit Public Library.

FIGURE 1-8. Michigan Carbon Works factory grounds in Detroit, Michigan, photograph, circa 1895, Detroit Public Library.

The tower of bones appears nearly thirty feet high. The tidiness of the pile, its scale, and the choice of the photographer to shoot from below draws attention to the aesthetic details of the pile. But we are not looking at a photograph intended to be strictly an aesthetic object, such as a landscape photograph. There is an obvious economic purpose to this image, which advertises the fortune and capacity of this corporation. Yet, overlapping with the economic value of this image is the aesthetic sense of formal composition, the sublime sense of nearly overwhelming scale, and the dramatic spectacle of so much death.

After taking in the tower of skulls, the viewer's eyes move from the symmetrical repetitiveness of the stacked bones to the two men, one at the foot and one at the top of the pile. The men in this photograph are probably Deming Jarves and William D. Hooper, the founders of the company. Both men stand with one foot propped on the top of a bison skull. The meaning of this stance is unmistakable; they are posing in the familiar gesture of the "trophy shot," typified by the hunter standing on or next to his kill. Matthew Brower, in *Developing Animals*, explicates how the same techniques and strategies used in hunting were employed in photographing animals. The genre of the trophy shot follows a consistent visual logic in displaying "the dead animal as testimony to the hunter's prowess."[30] The trophy shot epitomizes the convergence of photography and death, camera and gun. Brower adds that trophy shots function as an appropriation of the animal into a symbolic code that demonstrates the transfer of power from the animal to the human. "The trophy photograph links the photographer as a predator or hunter to the animal subject; it situates the photographer as the consumer of photography's symbolic kills" (81).

The gesture of standing over a killed animal can be understood as a singular image of what Alfred Crosby calls "ecological imperialism,"[31] the domination of native flora and fauna concomitant with the domination of Indigenous peoples by settler colonial enterprise, followed by the displacing of native ecologies with plants and animals from Europe (the bison will be replaced by European cattle and prairie grass will give way to wheat, cotton, and corn). In the scene of the Detroit rendering plant, the men theatrically display an attitude of conquest, taking obvious efforts to look confident and triumphant atop an enormous wall of bison skulls numbering in the tens of thousands. However, some aspects complicate defining this image as a trophy shot, since the men pictured were not involved in hunting or killing the animals. These bones were bought, scavenged from the remains of the decimation of the bison in the previous decade. Yet the triumphant gesture of the men is unmistakable; the trophy here is the incredible proportions of the bone pile ready to be converted into animal capital.

Since the photographer of the Michigan Carbon Works is unknown, the intended audience and destination of this photograph is hard to discern. Did the president of the company ask for and keep the photograph, or did the photographer take the initiative, perhaps to garner a reputation for his studio? Is this photograph taken from a corporate point of view or a citizen's view documenting a moment of organized human labor in modernizing the animal rendering industry in America? If the latter, the image is an early example of the social realist aesthetic of documenting the transformations of labor in connection to animal rendering processes that will reach its apotheosis in Upton Sinclair's *The Jungle*. The negative for these images was obtained by a collector of photographs of Michigan and, eventually, donated to the archives of the Detroit Public Library. The collection at the library gathers evidence of Michigan's manufacturing prowess at the turn of the century, which positioned the city to be a prime location for the nascent automotive industry that flourished in subsequent decades.

Upon further inspection of the images, one notices evidence of the importance of the size of the rendering plant, with large smokestacks in the background, warehouses in the distance, and a glimpse of train tracks in the foreground. The colossal bone pile overwhelms the foreground of the image and makes the factory buildings appear smaller, although this kind of visual trickery reminds the viewer that the factory grounds must be massive, indeed, to be able to process all the bones. These trappings of industrial modernity—industries specializing in large-scale operations, trains and transport networks, making mass-market commodities—date the image as contemporary with the burgeoning development of "Fordist" assembly-line manufacturing, a transformation in the organization of labor and machinery that Shukin shows would not have happened without the prosperous nexus of capital with the material and symbolic power of animals.

The haunting power of this image is due in part to its demonstrations of normalcy and marketplace efficiency. The bones are neatly stacked not out of a sense of care for the bison's memory but to make the pile more stable, countable, and ready for processing in the furnace. The comportment of the men suggests satisfaction and managerial assurance. They do not show expressions of astonishment or dismay. Their gestures confirm they are standing on top of a heap of capital, not a mountain of death. Their pose flouts contemplation of the overwhelming loss of life that surrounds them. When we look at the photograph today, it is the very calmness and lack of disturbance characterizing the scene that becomes so disturbing. The photograph is dizzying with the scale of death yet also depicts business as usual; it is both jaw-droppingly shocking and exemplary of unremarkable everyday practices of rendering bones into capital. The image shows an ordinary day at the factory and the extraordinary, epochal moment of the final stages of a species' demise. The jarring contrast between the extinction event of the bison and capitalist normalcy elicits a dissonant and anguished gaze. The image is celebratory and haunting, oblivious and violent, mundane and overpowering, junkyard and mass grave. The photograph is a perfect example of the oft-cited observation made by Walter Benjamin that every document of civilization is also a document of barbarism. The men in the photograph are standing on what is probably the largest pile of bones, animal or human, that had ever been photographed up to that point.

The Michigan Carbon Works photographs cannot be understood without the context of previous bison imagery and rhetoric that intertwined depictions of the massive bison herds with the vision of the last bison.[32] The trophy shot iconography in these images combines with the visual history of the extinction shot. Extinction shots are the pictorial evidence of last animals and create a visual culture around the last animal figure. The extinction shot typically shows a single animal that is construed as both an individual and an iconic representative of its species. When a photograph becomes saturated by its identification with extinction, the image becomes a locus where expressions of care, grief, regret, blame, and salvation reach extreme intensity.

The importance of recording the last animal with visual technology, either by photograph or by film, became an identifiable object of desire as animal photography developed at the end of the nineteenth century, at the same time a scientific interest in identifying at-risk species began to take hold. The use of the camera to document last animals, whether in the wild or in captivity, became central to detailing and authenticating the lives and numbers of last animals for scientific purposes, but also for spreading awareness to the wider public about the animal's imminent loss. Whether or not the photographer is conscious of the reality of extinction is not a primary factor—it is the camera itself, with its mechanical, impassive "eye," that indexes a historical moment of last life and "sees" the extinction event.

More recently, in the last few decades, the practice of last animal photography and the genre of the extinction shot have proliferated, supplying a plethora of visual references for the rapid loss of biodiversity and animal populations in the Anthropocene. Photographs of extinct or endangered animals in recent decades explicitly aim to work against the spectatorship of *animalis nullius* perpetuated in earlier images of animal finitude. However, even these more recent images of animals facing extinction divulge a conflicted legacy of looking at animal suffering and disappearance. The images from the Michigan Carbon Works are another iteration of the gaze of *animalis nullius*, and, as such, it is necessary to observe how this colonial gaze works while also situating these images within a broader reflection on what it means to look at extinct animals from the historical vantage point of today.

Extinction shots, which contribute to the wider repertoire of the visual culture of ways of looking at animals, are unique to the medium of photography in a similar way as trophy shots. Understanding these stylized images of animal mortality requires a re-examination of the often-remarked fundamental connection between photography and death. "Ever since cameras were invented in 1839," Susan Sontag remarks, "photography has kept company with death."[33] We have already seen how the camera repeatedly overlaps with the gun, especially on the plains of North America. Both camera and gun served as frontier technologies used to "tame" and "clear" and "close" the Western frontier. Sontag also discusses how photography has played a key role in all wars since its invention, serving as reconnaissance, propaganda, reportage, and documentary of the realities of violence (66). Moreover, at the very formalist level, the photographic image itself, regardless of its subject matter, is caught up in the relationship of photography with death. Photographs fix a moment in time that is ever-receding to whomever views the image. This fixed exhibition of a past cannot but evoke in the viewer a sense of time lost, the fading of history, and the perishability of what once was.

André Bazin, in "The Ontology of the Photographic Image," associates photography with the long-standing belief (dating as far back as ancient Egypt) in the power of the plastic arts to effect "the preservation of life by a representation of life."[34] Photographs have the ontological power to fix a moment in the life of someone or something and make it so that the life-image endures in perpetuity. The price for this representation of life, however, requires that the photographic medium itself takes on the preservative work of death. Bazin states that "photography does not create eternity, as art does, it embalms time, rescuing it simply from its proper corruption" (14). Embalming is an ambivalent state between life and death. The fixed image of the photograph, an event that once happened in time and then is taken out of the flow of time, forces the viewer to ponder the recession of the event from the point of view of the present. Both loss as well as preservation are fundamental to the ontology of the image.[35] Sontag comments that it is an inherent property of all photographs to provide an image of loss itself, and, as a result, photography enlarges the "iconography of mystery, mortality, [and] transience."[36]

Roland Barthes goes as far as to describe photographers as "agents of death."[37] In Barthes articulation of what he calls the "image's finitude" (90), the ontology of the photographic image "is a kind of primitive theatre, a kind of *Tableau Vivant*, a figuration of the motionless and made-up face beneath which we see the dead" (32). Every photograph presents a kind of stationary pose or *tableau vivant* of someone or something that has been extricated from the flow of time. Every photograph is, then, an image of lastness and lostness. For Barthes, portraits have an especially haunting poignancy because every photograph of a person implicitly says: this person is going to die. "Whether or not the subject is already dead, every photograph is this catastrophe" (96).

Extinction shots are extreme manifestations of this catastrophe implicit in the ontology of photography. However, the general condition of the entanglement of the photograph with loss and mortality is not the same as actual images of death of a human or nonhuman animal. Understanding extinction shots requires grappling with the finitude implicit in the ontology of every photographic image and the specific material and historical processes that led to any particular animal facing an extinction event. Extinction photographs make visible the painful intimacies of the photographic image with death and the passing away of the animal's world; they also reveal, at the same time, evidence of the role of specific actors, attitudes toward the disposability of animal lives, and the economies centered on the material remains of animals that lead to the radical decline in numbers of a species.

There is an additional formalist property of traumatic photographs that bears special attention in coming to terms with the utterly compelling exposure of the extinction shot. Peggy Phelan, in her discussion of the work of trauma elicited by photographs of atrocity, examines how such photographs activate a psychoanalytic process in which seeing and loss of vision are tied together. Traumatic events have the effect of fixating memory on the traumatic experience while also refusing to the mind full cognitive access to that memory. In Phelan's analysis, photographs implicated in trauma and atrocity "both document and create blind spots"[38] in that "these photographs expose the essential blindness that constitutes the act of seeing as such" (55). Traumatic photographs can bring together the fixation of vision and the destruction of the visible. Such photographs couple the unseeable and the act of seeing in a kind of short-circuit effect that is felt as the "flash" of recognition and the disturbance of cognition upon looking at the image. This flash or short circuit is a breech or wound in the psyche caused by a violent stimulus that hits directly on the mind with no effective mitigation.

To look at a traumatic photograph, such as an extinction shot, is to glimpse in an instant an image that fundamentally opens one's eyes yet also exposes the witness to something too painful to take in visually. To be sure, the psychological effects of looking at images of trauma are not at all commensurate with the violence felt by those individuals whose lives are depicted in the image. Such photographs do, however, solicit further reflection on what it means to look at such images in a responsive and responsible way, which includes becoming aware of how the viewer's gaze is implicated in a broader visual culture that has made it possible to observe and document extinction.

Photographs of extinction plunge the viewer into this intensely dissonant cognitive process whereby a glimpse of the end of a species produces a psychic shock that both stimulates and incapacitates the gaze. There is a short circuiting even in the phrase "photographing extinction" or "thinking about extinction"—photographs of extinction expose the traumatic end of the visible (the last gaze of a life form) just as thinking about extinction reaches the traumatic limit ends of thought. All photographs are indexes of a real moment; in the case of the extinction shot, however, there is a short circuit between recording reality and the loss of reality for the species. Extinction photographs make visible how the loss of a life form leaves a puncture, a wound, or a hole in nature. These photographs freeze in time the end of the animal's temporality.

In Barthes's analysis of the ontological relationship of photography and death, he finds that every image states "that-has-been."[39] The exposition of "has-been" or "this was" characterizes the existential formalism of photography. Extinction photographs, in which this existential formalist quality is itself put into crisis, are images saturated with "no-longer-there-ness." The effacement of "there-ness" of a species, more radical than just the inevitable recession over time of all images, is intensified in the convergence of the ontological loss evident in the photographic image and the loss of the animal. Extinction photographs are visible evidence of the "gone-ness" and futurelessness of the animal, its way of living in the world, and its contribution to biodiverse ecological communities. The compulsion to take extinction shots in present times, to photograph all endangered species, testifies to the special power of photographs to document and make observable the "gone-ness" of life. Such images of "gone-ness" and "therelessness" serve an evidentiary as well as an aesthetic function that calls for an ethics of looking at animals that refuses predatoriness or the instrumentalizing gaze of animal capital. The photographic archive of last animals over the past few centuries serves as a haunting reminder of what the world was once like with these animals in it.

These same images are key inspirations for the possible return of these animals. For example, recent attempts to de-extinct the quagga, Tasmanian tiger, and passenger pigeon have relied heavily on drawing attention and desire for such efforts by circulating last animal images. In a number of cases, the scientific work of reviving these species is based on trying to get the animal to appear like it did in photographic images taken while the species was alive. These images have become central to several recent de-extinction efforts to the extent that recreating the animal in its visual likeness, such that the revived species looks visually identical to its ancestors, is proving to be as important as fully recovering the extinct species' genomic profile and ecological niche (an issue discussed further in chapter six).

Because the "last animal" today draws so much attention—even more so in a time of rising extinction rates—it is now expected that an extensive documentation, especially at the visual level, will occur for endangered animals. Book projects collecting the photographic record of endangered and extinct animals, such as Douglas Adams and Mark Carwardine's *Last Chance to See*, Errol Fuller's *Lost Animal: Extinction and the Photographic Record*, and Joel Sartore's *Rare: Portraits of America's Endangered Species*, continue to show today how last animals, technologies of visibility, and conflicted affective responses in viewing such lastness are intertwined in the development of the extinction shot.[40]

Photographers of last animals today often find they need to aestheticize the image of the animal to avoid any hint of the animal's commodification or subjection. In Sartore's photographic series of critically endangered species, for example, he supplies a white or black backdrop for each animal, which makes the animal starkly visible but also emphasizes the animal's isolation and displacement from its habitat. The animal becomes more iconic and acutely sympathetic in the image that stages a "last" pose. The emphasis on physique and gesture together with the flat background reference the visual history of portraiture, in particular the "last" imagery of Christ. This raises the concern in Sartore's images that it can feel like the animal is being enticed to pose for its own extinction shot. Such contemporary extinction photographs have to tread the line carefully between framing the lastness of a species as presented for respectful contemplation and a framing that decontextualizes and too readily evokes a kind of animal canonization.

The photographs of the Michigan Carbon Works defy the desensitization that typically follows from the wearing out of the power of an image over time. These photographs haunt now more than ever, even though the bison did not go extinct and have, in fact, returned, albeit in much diminished numbers, to the North American plains (a situation discussed further in chapter five). The towering pile of skulls, the overwhelming presentation of mass death, cannot help but create associative flashes in the mind toward other images of atrocity. A sense of painful commonality across the species divide is referenced as the Carbon Works photographs enter into association with other photographs of destroyed bodies and worlds undone.

Sontag's insistence on looking at photographs of inhuman acts while asking humanist questions about such looking did not prompt her to wonder what happens when viewers regard the pain of animals. Her observation that "photographs echo photographs"[41] can apply to how such photographs of painful animal ends can become vivid reference points for human ends. The Carbon Works bone pile becomes more legible and visible by association with subsequent photographic evidence of mass death. It seems unavoidable that images of the mass devastation of bison life should be connected to images of devastated human life that includes the concentration camps of World War II, the Killing Fields strewn with Cambodian bodies executed by the Khmer Rouge from 1975 to 1979, and the mass slaughters during the Rwandan Civil War in 1995.

The instant objection to associating animal photographs with human photographs of atrocity—specifying that an animal rendering plant is not at all at the same level as instances of crimes against humanity—is an appropriate response but does not quite mitigate the evocative resonance between images of mass death, both human and animal, that expose the viewer to witnessing the extreme limits ends of life in an extinction event. Instead of reasserting the anthropocentric gaze yet again to insist that animal finitude has no comparison whatsoever to human finitude—as if anthropocentrism itself had not already been throttled by images of humanity's undoing in these horrific genocidal instances—the viewing of images that make us expressly aware of species extinction calls for a different ethics of seeing. It is necessary to develop a visual knowledge that can grasp the specific historical contexts and agential acts involved in each extinction event while also recognizing how such images present the inverse of a shared community of life on Earth. These extinction shots, each in their own way, make visible the unraveling of biodiversity and the commons of life. They frame the unsharing of the planet by making natural and cultural diminishment a viewable phenomenon.

On August 29, 1911, an Indigenous man, later given the name Ishi, approached a slaughterhouse on the outskirts of the small town of Oroville in Northern California. He was emaciated and alone. Workers at the slaughterhouse contacted the sheriff, who took him to the jailhouse, not for any infraction but because the sheriff did not know what else to do. Several local Indigenous persons from communities in the surrounding areas were brought in to speak with Ishi, but they had not a single word in common. The town's newspaper published an article about his haggard appearance and speculated about his origins, and the article was shared on a press wire and reprinted in other newspapers in California, including in San Francisco, where Alfred Kroeber, professor of anthropology at the University of California, Berkeley, read the account. Kroeber offered to take the Indigenous man into his care, and Ishi would spend the rest of his life dwelling in a museum run by Kroeber, not as an exhibit but as a paid custodian.

Ishi knew he was the last living member of the Yahi people, who had been steadily killed by white settlers. Upon moving to San Francisco, Ishi found himself embraced by whites, and was hailed as "the last wild Indian in North America,"[1] in the phrasing of the biography of Ishi written by Theodora Kroeber, Alfred Kroeber's second wife.

Theodora Kroeber's retelling of Ishi's life, and the subsequent stories and accounts of his experiences, employ troubled references to firsts and lasts that have been developed in tandem with conceptualizations of extinction. Kroeber did not use the word "vanishing" in her biography, wary of the denigrating associations of that particular term, but she did retain a number of similar terms to indicate Ishi's "lastness." Her writings on Ishi's life and the Yahi, who are one of five subgroups historically connected to the Yana Peoples, are explicitly retold as extinction events: "We *know* the history of the Yana only at that last moment before its extinction, when it was become meagre and flickering" (198). Kroeber reflects on the painful admission that her writing conveys to the reader the richness of Yana life and culture only after its destruction.

However, although Ishi was the last of the Yahi people, he was not the last person of Yana heritage. Describing him as the "last wild Indian" played on romantic stereotypes of disappearing indigeneity that Kroeber criticized in some respects but amplified, as well, in her own way. Ishi's story is not the wholesale end of a certain Indigenousness; rather, his situation emblematizes how Indigenous peoples across the Americas adapted themselves after contact and under genocidal circumstances, by way of acts including cultural sharing, resistance, and traditional continuations. Ishi's life, set in the midst of long-standing Indigenous practices of resistance and resurgence to counter the "ecocide-genocide nexus," shows a redefinition of terms of extinction away from searching for and expecting final vanishings. Instead, Ishi's experience associates extinction events with making new forms of cultural transmission, storytelling, and fostering decolonial relations, even while testifying to the violent undoing of his people.

Ishi welcomed the recording and sharing of his language and culture, and his commitment to sharing Yahi culture is part of his story of living through the extinction event perpetrated upon his people. Anishinaabe writer and scholar Gerald Vizenor, whose several writings on Ishi have played a fundamental role in extricating his story from what Vizenor calls "terminal metaphors,"[2] insists on identifying with precision what Ishi's lastness entails. "You were not the last memories of the tribes, but you were the last to land in a museum as the simulations of a vanishing pose in photographs."[3] Indeed, since Ishi's death and the destruction of the Yahi community took place in a historical period that saw a distinct rise in the reclamation of rights and powers of self-determination for Indigenous peoples in California, Ishi's experience did not represent a "closing chapter" of California Native life but, rather, a "retribalization of California,"[4] in the phrasing of Orin Starn.

Is extinction, then, the right term to describe Ishi's experience and the eradication of the Yahi by California settlers? In Ishi's time, the political leadership of California talked openly of sanctioning the extinction of Native Americans. The first governor of California, Peter H. Burnett, declared in 1851, "A war of extermination will continue to be waged between the races, until the Indian race becomes extinct."[5] In the nineteenth century, both in common use and biological science, the term *extinction* applied to the decimation of species as well as the eradication or forced assimilation of races and cultures. Charles Darwin, in *The Descent of Man*, expounded upon "the partial or complete extinction of races"[6] and assumed that "extinction" applied to situations without distinction to biological death, cultural destruction, or colonization of communities and peoples. According to Darwin, "Extinction follows chiefly from the competition of tribe with tribe, and race with race" (189). Darwin declared such instances "extinction" when one culture or peoples dominated or displaced another either by absorbing or destroying the other. Darwin did recognize that multiple factors and causes contributed to any one extinction event, yet confidently declared that the demise of "races of man" followed along the same conceptual lines as the extinction of animals: "although the gradual decrease and ultimate extinction of the races of man is a highly complex problem, depending on many causes which differ in different places and at different times; it is the same problem as that presented by the extinction of one of the higher animals.... The New Zealander seems conscious of this parallelism, for he compares his future fate with that of the native rat now almost exterminated by the European rat" (199).

Theodora Kroeber expressed her hesitation with biological definitions of extinction in the situation of Ishi, and she wrote that "the Yahi were not, after all, a 'species.' They were a macrocosmic nation victimized by the common killers: invasion, war, famine, and intolerance" (95). Kroeber rejected Darwinian-era rationalizations of the eradication of Indigenous peoples because such terms promoted the view of a naturalized biological process rather than recognizing specific human causal agents who perpetuated oppressive and politically heinous acts.

Extinction under conditions of natural selection can happen for a number of biological and climatic reasons, including competition for food and habitats among similar species in geographically restricted spaces. Genocide happens when resources and lands are stolen, and peoples are dispossessed and disposed of in explicit acts of subjection and violence that have nothing to do with natural selection. Furthermore, *extinction* specifically denotes the end of a species, and so it would seem inaccurate to describe anything less than the total end of all humanity. Yet the terms of extinction and what can be called an "extinctionary logic" also have circulated in a more diffuse sense throughout the nineteenth and twentieth centuries in acts of genocidal and exterminationist violence.

Although Kroeber does not use the term *genocide*, coined by Raphael Lemkin in 1944, her narration of the killing of Yahi peoples fits the definition. The terms of extinction, which include natural and artificial histories, and genocide, which is always political, require dedicated and continual parsing in specific times and places. However, in the situations of the genocide of Indigenous peoples in the Americas and the genocidal agenda of the Third Reich (discussed in chapter four), the perpetrators forged an extinctionary politics that often blended the terms in ways that reinforced attitudes of perceived natural and political dominance.

One important distinction between genocide and extinction made recently by Indigenous scholars addresses the issue of how settler states have wielded the designation of the finality of extinction to dismiss Indigenous grievances and land claims. Ward Churchill and Winona LaDuke describe how the term *extinction* as applied to Indigenous communities has been used by U.S. and Canadian governments to terminate tribal recognition and abrogate treaty and title rights. "By-and-large, 'extinction' is and has always been more a classification bestowed for the administrative convenience of the settler-states than a description of physical or even cultural reality," Churchill and LaDuke remark.[7] "Although both the United States and Canada officially maintain that genocide has never been perpetrated against the Indigenous peoples within their borders, both have been equally prone to claim validation of their title to native lands on the basis that 'group extinction' has run its course in a number of cases. Where there are no survivors or descendants of preinvasion populations, the argument goes, there can be no question of continuing aboriginal title" (53).

Extinction as a term borrowed from biological and natural sciences also makes for a problematic referent when ported into cultural linguistics to describe the disappearance of a spoken language or cultural practice. In their essay "Surviving the Sixth Extinction," Indigenous linguists Daryl Baldwin, Margaret Noodin, and Bernard C. Perley argue that extinction is an "invasive concept"[8] to describe a language not actively spoken because the biological denotations of the term indicate permanent and irretrievable loss, whereas languages with few or no current speakers can be revivified or reappear in other ways. Instead of applying the metaphor of extinction to language loss, which often has motivated linguists and anthropologists to "save" endangered languages without critically reflecting on how this effort promotes the expertise of the linguist disconnected from present and future Indigenous language users, they recommend considering such languages as "asleep." This metaphor of a sleeping language is more consonant with Indigenous-held views of the importance of maintaining a relationship between speakers, language, and local lands that requires replenishing over generations: "Perhaps chasing lost speakers has never been relevant and instead we should be nurturing the first speakers of the next version of the language" (219).

This chapter examines how discourses of extinction and eradication circulated in the narratives of Ishi's "lastness" and were interpreted as representative of the destruction of Indigenous peoples in North America. Reflecting on Ishi's witnessing of the destruction of his community requires addressing both the decimation of the Yahi Peoples and the ongoing legacies of Ishi and Yahi knowledge, including up to today. Ishi, who never revealed his name once he left his ancestral lands, did not want to be unknown and did not want to be fully known. Still, Ishi found ways to develop forms of Indigenous affirmation in the context of the violent end of his people. Gerald Vizenor has written powerfully of the need to understand Ishi's life not as a tragic closure but, rather, as an instance of "survivance," a portmanteau word that combines survival and resistance and carries within a sense of possible futurity.

Following Vizenor's emphasis on Ishi's enduring metamorphoses, this chapter examines the extinctionary acts and policies that brought Ishi to San Francisco and how Ishi found ways to tell his story of what it is like to live through an experience of extreme violence. Ishi navigated his way through everyday acts of survivance and endurance amid settler-colonial fantasies concerning Indigenous origins and ends. Cast out from his ancestral lands and into an urban Indigenous experience within the span of a few days, Ishi became an advocate for preserving and recording Yahi lifeways while positioning himself as rejecting the premodern/modern dichotomies used to characterize his experiences.

Under extinctionary conditions, Ishi tells old stories and lives new ones, transforming himself into a bearer and sharer of traditional Yahi knowledge in the context of bustling capitalist San Francisco. Ishi saw whites as potential friends and, possibly, concerned allies but also knowingly implicated in the history of Indigenous dispossession and destruction. Ishi's commitment to maintaining acts of friendship and hospitality even when suffering such extreme loss stunned Alfred Kroeber and many others who connected with Ishi. Ishi's generosity was not a compensation or denial of his traumas but part of his teaching to others from his Indigenous experiences of what it means to live with strength through an extinctionary event. The conjunction of Ishi's tremendous loss and depth of graciousness had a transformative effect on Kroeber himself, who later said of Ishi, "He was the most patient man I ever knew. I mean he had mastered the philosophy of patience, without a trace either of self-pity, or of bitterness to dull the purity of his cheerful enduringness."[9]

Alfred Kroeber wrote on several occasions about what he called "the last 'uncontaminated' aboriginal."[10] Indigenous historian Jean M. O'Brien, in *Firsting and Lasting: Writing Indians Out of Existence in New England*, examines how settler colonial tropes and doctrines of firstness and lastness already were circulating in the early decades of the founding of the United States as a means to dispossess and discredit the continued existence of Indigenous peoples. O'Brien discusses how discourses of "firstness" predicated on "temporalities of race"[11] followed from colonists' claims to discover and found civilization in the New World, thus legitimizing the prosperities that ensued from such theft. A similar claim to "lasts" along racial lines posited Indigenous peoples as fated to perish or become assimilated, hence disappearing in another way because they would lack the blood "purity" that constituted true Indianness. O'Brien points to how firsting and lasting followed from settler dominance of the terms of modern life, since "these scripts inculcated particular stories about the Indian past, present, and future into their audiences. The collective story these texts told insisted that non-Indians held exclusive sway over modernity, denied modernity to Indians, and in the process created a narrative of Indian extinction that has stubbornly remained in the consciousness and unconsciousness of Americans" (xiii).

It is crucial, then, to comprehend how references to Indigenous lasts circulate within the context of the history of thinking about extinction, and in narratives of Ishi more specifically, without perpetuating the mythologies and racial hierarchies embedded in the period. Understanding the history of concepts and practices of extinction should not reproduce the essentializing and naturalizing claims of "tragic" existentiality attached to intentionally destroyed Indigenous communities and lifeways. As Daniel Heath Justice (Cherokee Nation) remarks, "Having a clear and unromantic perspective about the many challenges that face Indigenous peoples is not the same thing as seeing those challenges as an innate expression of *our very nature*."[12]

Theodora Kroeber's biography saturates her account of Ishi's life with melancholic tones, and while it is appropriate to read her work as a somber tribute to Ishi's experiences, her writing encouraged readers to expand these melancholic affects as characteristic of the last and lost condition of indigeneity as such. These affects and metaphors of lasting contributed to a de-futuring of indigeneity. Re-reading her account through a critical reflection on the concepts of extinction used at the time involves being accountable to the history of genocidal acts committed upon Indigenous communities while decoupling these events from what Theodora Kroeber called the "duly fatalistic"[13] disposition of the Yahi People.

Alfred Kroeber's encounter with Ishi and the emotional effects of Ishi's friendship contributed to Kroeber's own questioning of the discipline of anthropology. Shortly after Ishi's death, Kroeber abruptly put his anthropological career on hiatus and turned toward psychoanalysis, first as a patient and then as a practitioner. As Kroeber increasingly engaged with Freudian methods and ideas, he also sought to question Freud's investment in ahistorical anthropological figurations of the "savage" and "primitive" mental states in psychoanalysis.

In fact, Kroeber's own anthropological and psychological writings push back against Freud's fantasies of Indigenous origins and ends. Kroeber hoped to align some of Freud's psychoanalytic speculations on culture and the death drive with Kroeber's own efforts to synthesize empirical work in anthropology with more theoretical analysis of cultures. Freud's own tendentious speculations on Indigenous life and anthropology already are widely known. What is less discussed is how Freud's arguments on trauma and the death drive drew from discourses of indigeneity and extinction that Freud then implicated in the origins and ends of the psyche.

Psychoanalysis requires a theory of the origins and ends of the psyche as fundamental to its own disciplinary self-analysis. Freud found it necessary to place a theory of origins and ends that borrowed heavily from colonial discourses of firsts and lasts at the heart of psychoanalysis. While these terms become self-deconstructing, as Derrida has shown,[14] they raise the issue of how to re-conceptualize the finitude of the psyche as distinct from the extinctionary histories with which these psychoanalytical terms are intertwined. Such analysis also leads to questioning how Freud's arguments on the death drive have more recently been taken up as exemplary for thinking about extinction today.

## Learning from Ishi

As a settler scholar, I examine some of the ethical entanglements involved in the way Ishi's life has been situated in settler colonial terms of sympathy and nostalgia for last and lost lifeways. I first became aware of Ishi's life as an elementary student in Contra Costa County, California, situated on the traditional territory of the Ohlone Peoples. Ishi's story was presented to our class with a short video documentary of his life, *Ishi in Two Worlds* (directed by Richard C. Tomkins in consultation with Theodora Kroeber), made in 1967 and intended for educational use. We had a curriculum unit called "Indian days" that briefly covered local Indigenous history narrated in a textbook on the history of California statehood. Our school abutted a grassy and hilly open space area, and we ventured outside where we were taught how to craft arrowheads, make acorn mush, and weave small baskets from tall grasses. We did not meet or hear from living Indigenous peoples. Many California students have a vague and ghostly recognition of someone named Ishi, whose story briefly appeared in school curricula or local cultural history museums. Items including arrows, clothing, and fishing tools made by Ishi and others from his community are on permanent display in the Phoebe A. Hearst Museum (now in Berkeley) and the California State Indian Museum in Sacramento. Ishi is likely the only local Indigenous person many settler Californians "know" by name.

Much of what is known today of Ishi's story comes from Theodora Kroeber's book *Ishi in Two Worlds: A Biography of the Last Wild Indian in North America*, published forty-five years after his death in 1916. Kroeber did not meet Ishi in person and compiled her book with the aid of notes and recollections from Alfred and from other testimonies and documents. The current print edition of Kroeber's biography comes with the back cover statement "over one million copies sold," a reminder that her account of Ishi has now passed through several generations of readers. Kroeber thought the popularity of Ishi's story lay in what she saw as Ishi's humanism in the face of horror: "Ishi's human appeal seems to have been a universal one, his history more interesting and bizarre in the face of his friendliness and a characteristic quiet élan" (171). She stressed this "universal" appeal in another statement: "Howsoever one touches on Ishi, the touch rewards. It illuminates the way."[15] Yet this inscription of Ishi into the domain of an almost mystical power to "reward" evades confronting the reader with any sense of the ethical responsibilities that non-Indigenous narrators and audiences have toward stories that are not their own.

The composition of the biography mirror's Kroeber's "two worlds" motif by splitting the book into two parts, the first covering the history of the Yahi People and Ishi's life in his ancestral lands, and the second part covering Ishi's life in San Francisco. The split narrative structure of the book also presents in its form the view that American policy toward Indigenous peoples has shifted from the nineteenth-century extinctionary practices of military conquest and elimination to doctrines of assimilation and sentimental attachments to Nativeness in the twentieth century. Kroeber's biography was one of the first popular works to provide a counter-history of California settlement. The book displaced the widely promoted California nostalgia for frontier towns and gold rush history as it detailed the numerous massacres of Indigenous peoples across the region prior to and after admission of California to the United States.

Nearly half of Kroeber's book tells of the Yahi People pre-contact and the several horrific incidences of murder perpetrated against the community in the nineteenth century by white settlers with impunity from government authorities. There also is a hint of a Berkeley-based countercultural ethos in her book that puts Ishi in the position of offering a critique of the vanities of modern American life. She situates Ishi between pre-contact California innocence and American national guilt, and suggests to readers that only a rededication to "panhumanity" (230) would heal such a rift going forward. Kroeber treats Ishi as someone who was traumatized twice, by being the last of his "Stone Age" people and the first of his people to live in highly modernized and alienated conditions. Her book also borrows heavily from a 1950s American psychoanalytic lexicon to describe and interpret Ishi's story in terms of individual and collective psychological experiences. Kroeber saw in Ishi's lastness something profound about the traumas of crossing "certain physical and psychic limits" (10) and the need to understand the psychological impact of human extinction events across cultures and time periods.

*Ishi in Two Worlds* is also, in part, a tribute to Alfred and his anthropological work, although Theodora Kroeber did not write as an anthropologist and purposefully blended genres of biography, ethnographic study, local history, and theorizing about cultural difference and civilizational ethics. Her writing is more intimate and emotionally effusive than an anthropological study and explicitly aims for a popular rather than scholarly audience. Most distinctly, Theodora Kroeber dwells extensively on Ishi's personal experiences and feelings. By contrast, Alfred Kroeber, in a 1915 essay laying out his methodological principles, declared that "the personal or individual has no value save as illustration."[16] Although Theodora Kroeber does not expressly critique the field of anthropology or point out its shortcomings, her aesthetic choices, such as her use of mixed genres and attention to domestic and everyday experiences (perhaps also indicative of Kroeber's keenness toward nascent feminist methodologies in biographical writing), serve to distance her work from academic ethnography. However, the technique of intimate storytelling that places the reader at close quarters with Ishi also follows from what Shari Huhndorf describes as the trend in the twentieth century of settler fantasies of "going native." The attraction of immersion in indigeneity that Kroeber's book offers, Huhndorf writes, participates in "idealizing and emulating the primitive, modernity's other, comprised in part a form of escapism from the tumultuous modern world. Consistently, throughout the twentieth century, going native has thus been most widespread during moments of social crisis, moments that give rise to collective doubts about the nature of progress and its attendant values and practices."[17]

With the book's countercultural spirit of white nostalgia for indigeneity and its critique of unrelenting, aimless capitalism and American boosterism, Kroeber's biography helped contribute to the swooning mythologization of California as a land of sun-drenched peoples with "alternative" spirituality. Kroeber titled her first chapter "Copper-Colored People on a Golden Land," hinting at a utopian past of charmed people basking in warmth and plenty. Her work covered the whole psychoanalytic gamut from fantasies of childhood innocence (projected onto Indigenous others) to guilt-wracked maturity, passing through traumatic recognition of past violence, and culminating in the work of mourning to reach a sadder and wiser humanistic way of life. In a foreword to the 1961 edition, Lewis Gannett placed Kroeber's book in the context of Americans recovering from two world wars that have given "new poignancy to the phrase 'death of a race,' as well as to the conception that patterns of life other than our own may have validity and significance" (xxvi). The book brought together in extreme juxtaposition the desire for more inoffensive, alternative ways of coexisting while seeking to face up to the responsibility for recognizing extinction events and genocide at home.

## Contact as Extinction Discourse

The first appearance of Ishi in Kroeber's biography already places him in the position of crossing the threshold of "wild" to urban life. Emaciated and traumatized, Ishi stumbled into the town of Oroville, at which point he enters Kroeber's narration: "The story of Ishi begins for us in the early morning of the twenty-ninth day of August in the year 1911 and in the corral of a slaughterhouse. It begins with the sharp barking of dogs which roused the sleeping butchers. In the dawn light they saw a man at bay, crouching against the corral fence—Ishi" (3). Ishi first emerges enmeshed in a scene marked by the crossing of human and animal. He is caught at the back of a slaughterhouse, a place of animal death. Ishi is found "crouching" and the sheriff is called to take the "wild man" away (3). Ishi is immediately brought to the prison, but not to isolate him from public gawking, as many residents of Oroville come to visit, looking at him through the bars that are evocative of a zoo. They offer him food to see what he will eat and display various objects to see which ones will take his interest.

Kroeber does not dispel the animalesque and primitivist tones of this opening scene, as she repeats the word "wild" several times, a term that is supposed to indicate Ishi's pure Indianness (neither mixed breed nor assimilated into white customs), his fusion with nature, his utter lack of knowledge of modern civilization, and his ragged, apocalyptic appearance. Although Kroeber means such wildness to be a badge of distinction, the term evidently indicates otherness to modern civilization and distances and isolates Ishi from Kroeber's contemporary readers. Instead of foregrounding Ishi's intimate knowledge and relationship with animals regional to his ancestral lands, we find Ishi positioned into a different kind of animality that marks him as out of place and in a position of subjection. Kroeber simulates for the reader the experience of "first contact" with Ishi, as if following with Ishi his first steps out of the wilderness through the liminal human–animal space of the slaughterhouse (although much research since has shown Ishi certainly had multiple contacts with whites and Spanish-speaking settlers over the previous decades[18]).

Māori scholar Linda Tuhiwai Smith discusses how casting "authentic" indigeneity as "wild" is a colonial strategy that serves to essentialize cultural differences by basing definitions of indigenousness on dictums of pre-contact pristineness and the prohibition of adaptation or mixing. Smith remarks, "At the heart of such a view of authenticity is a belief that Indigenous cultures cannot change, cannot recreate themselves and still claim to be Indigenous. Nor can they be complicated, internally diverse or contradictory. Only the West has that privilege."[19]

Kroeber's prose mirrors for dramatic literary effect the language of the reports of those who first encountered Ishi. An article in the *San Francisco Examiner* of August 31, 1911, described the Yahi Peoples (also called "Deer Creek Indians") mixing terms of wildness and animality: "Wilder than other tribes, the Deer Creek Indians fled before the white man's approach—fled into mountain fastness where they lived as the beasts."[20] This article ends with the statement: "It is believed that the Indian captured is the only survivor of this band, and hence the least civilized man in America" (97). Although Alfred Kroeber was wary of the sensationalism around Ishi and wanted to protect him from becoming exploited as an item of entertainment and curiosity, he used similar language in an article that appeared a week later in the *San Francisco Examiner*. Kroeber, with his imprimatur as a professional anthropologist from a distinguished university, stated, "I can safely say we have the most uncivilized and uncontaminated man in the world to-day" (100).

Both Alfred and Theodora Kroeber were wary of denigrating terms that stereotyped Native peoples, yet they also drew heavily from the "wild man" figure, long seen as an uncanny, unruly, untamed double to civilized life.[21] References to the "uncontaminated" world of Ishi set him up for a fall into modernity, and scripted in advance the plot of the melancholic disappearance of "the wild" and the closing of the frontier that ideologically signifies the end of supposedly authentic indigeneity. Patrick Brantlinger points out that tropes of "the lament of the dying, often last aboriginal"[22] were common in the nineteenth-century literatures of recently formed settler colonial nation states. These terms echoed the rhetoric of progress and the adoption of evolutionary themes as indicative of the necessary "law" that primitive human societies must give way to more complexly organized and industrialized social systems riding on the forward march of history. Such tropes fed into what Brantlinger calls an "extinction discourse" (1) that "has served as a primary motivation for the funereal but very modern science of anthropology in its attempt to learn as much as possible about primitive societies and cultures before they vanish forever" (5).

The opening scene in Kroeber's book sets the intellectual and psychological schematics for Kroeber's story of the "two worlds" of the ancient aboriginal contrasted with the thoroughly modern. Because the dichotomy of two worlds follows along fixed binaries of ancient/new, past/future—suggesting that the shift to the latter marks the loss and erasure of the former—this motif implicitly affirms what David Wallace Adams points out as the common attitude in the nineteenth and early twentieth centuries that Indigenous peoples must choose either "civilization or annihilation."[23] Kroeber shifts between these two worlds at the will of her own narrative provenance. Her book reconstructs the traumatic extinction event of the Yahi People and then swings abruptly into the space of everyday San Francisco, seeming to replicate at a formal level the psychical blow Ishi faced in the forced loss of his world and his assumption of a second world. Although Kroeber left many indications in her book that she viewed modern life in San Francisco as convenient but prone to being shallow and wasteful, her positioning of Ishi precludes considerations of him as modern in his own way.

This anthropological trope of the fatal powers of white settler modernity parallels the biological trope that reappears consistently in extinction narratives that postulate the demise of a species due to its inability to find a place within globalized technological conditions. Ishi's appearance in San Francisco allows readers to relish in rediscovering their own fascinations with the bustle of the city and modern technological abundance, a point of view made possible only by indulging in Ishi's ancientness at the same time. As Douglas Sackman remarks, "When Ishi came to San Francisco he was seen as *the last* in order to bring into relief *the new*. Ancient and modern, an epochal ending and beginning, seemed to be present in the same place at the same time."[24] Kroeber and the reader enjoy the epistemological and narrative pleasure of flickering between these two worlds at will. The "two worlds" motif also reinforces the notion that, while genocide did, indeed, occur in destroying the original world of Native Americans, the modern space of San Francisco marked a different world of cosmopolitan reconciliation. Kroeber studiously avoids implicating Alfred Kroeber and Ishi's white friends—and, it should be said, the biography's reader—in any ongoing structures of Indigenous dispossession.

Though Theodora Kroeber's book was written in the late 1950s, at a time of growing Indigenous activism and cultural resistance against entrenched stereotypes of the noble savage, she continued to embed Ishi's story within the long-established tropes of the end of the "wild" but good-natured Indian, following unabashedly in the vein of *The Last of the Mohicans*. In emphasizing Ishi's lastness, Theodora again follows Alfred Kroeber's lead, who wrote a non-academic essay in 1912 with the title "Ishi the Last Aborigine: The Effects of Civilization on a Genuine Survivor of Stone Age Barbarism."[25] Even at the end of Kroeber's book, she pays tribute to Ishi in these same terms: "he was unique, a last man, the last man of his world, and his experience of sudden, lonely, and unmitigated change-over from the Stone Age to the Steel Age was also unique" (230). Kroeber positions Ishi both as extremely old and yet childlike, entering the Steel Age afresh.

The notion that Ishi came straight out of the Stone Age embellished the aura that coalesced around his lastness, but Theodora Kroeber knew Ishi had already come into contact with whites and material items of white civilization many times before 1911. Ira Jacknis observes that "Ishi had stopped being a 'stone age Indian' well before coming to live in San Francisco. He and his family were using arrow points made from glass, fish spear points from iron nails, and denim bags and hats, along with metal spoons and saws. These Yahi did their own collecting, finding objects abandoned or sometimes stealing them. As a kind of 'holocaust' survivor, Ishi lived his entire life in a state of cultural adaptation and change."[26]

The jailers in Oroville allowed two local photographers (John H. Hogan and Morris E. Phares) to take a series of pictures of Ishi the day after his discovery. One of Hogan's pictures appeared with the initial news articles that circulated across northern California papers, while others were printed as postcards to monetize on the popularity of the occasion.[27]

Several of the postcards show his exposed upper torso.[28] Theodora Kroeber reproduces one of the full profile photographs of Ishi in tattered clothing taken by Hogan (I refrain from reproducing the image as it was not taken with Ishi's consent). Ishi is positioned against a white backdrop, standing in the corner of a room in the jailhouse. The white background emphasizes Ishi's silhouette and feeble appearance. He is hunched with hands by his side, suggesting he has just been "caught" or "trapped" moments ago. He is shown without context, alone, and isolated. He is wearing a ragged sackcloth, not his native garb, and his hair is singed short as an act of mourning. As he stares straight into the camera, his appearance suggests he fits the pose of a last human figure who has been through an end-of-the-world event.

Hogan's photographs of Ishi are not meant to cast him in a romanticized and mythologized sentimentalism toward "vanishing" indigeneity; these photos appear, rather, as realizations of the intense and disturbing moment of "finding" or "capturing" Ishi. Hogan positions Ishi as starkly standing out, shorn of his dignity, visibly traumatized, with no relief from the settler gaze, his body fully exposed to the camera. The white background makes Ishi appear caught in a clinical light as if a kind of "specimen" of "last man." Ishi is shown solitary and without possessions, disconnected from his ancestral lands, and cast into a no-place. Throughout this ordeal, Ishi does not flinch. He returns the gaze of the viewer by looking directly forward with an expression of tensed readiness.

Shortly after arriving in San Francisco, Alfred Kroeber had Ishi photographed in a suit and tie, in both frontal and profile positions. Kroeber wanted to counteract the first images of the "wild man" shown in destitution and subjection. Subsequent to the taking of these images, Ishi would allow for several hundred photographs taken of him, many showing his skill in craftsmanship and practicing Yahi traditions (see figure 2-1).

The large record of photography of his life once he transitioned to San Francisco shows Ishi working to reclaim his image and his connectedness to his ancestral ways. They provide a counter-archive to these early photos of Ishi in distress—in effect, they are a refusal of the trope of the extinction shot associated with the first images taken of Ishi. After about a week of living in San Francisco, Kroeber put Ishi in a situation in which he was asked to disrobe his torso for a photograph. Theodora Kroeber notes that Ishi's refusal to undress constituted his first act of defiance. Ishi insisted on being photographed with his clothes on, rejecting the "wild man" pose. In quick time, Ishi began to understand his own photographic presence and visual identity, gaining a modernist sensibility of the mass

FIGURE 2-1. Ishi with fire making tools, photograph, 1914, The Bancroft Library, University of California, Berkeley.

culture phenomena of the reproducible image coinciding with his own Indigenous viewpoint recontextualized in urban and performative settings. Theodora Kroeber writes: "Ishi was photographed so frequently and so variously that he became expert on matters of lighting, posing, and exposure" (171).

Brian Hochman discerns how the ethnographic consciousness at the turn of the century developed in connection to new technological media, such that "new media like photography, phonography, and cinema provided tools to popularize the discipline of anthropology in the United States. . . . In short, the origins of modern media in the United States are distinctly ethnographic."[29] The way Ishi's story is told through first and last framings, in discourses of origins and extinctions, and the way Ishi participates in his own storytelling traditions are mediated by these modernist ethnographic technologies. Although Ishi would not find himself in full control of the camera machine in these photographs, he declined at times to have some images taken, instead welcoming photographs of him demonstrating archery, using the fire drill, crafting his tools, and telling stories. While standing as witness for the destruction of his people, Ishi refused to stand for vanishing poses. In photographs as well as wax cylinder recordings, Ishi willingly participated in the mechanical media of modernity that allowed him to make a recreation of everyday Yahi life.

## Ishi's Worlds

After introducing Ishi's "contact" scene in Oroville, Theodora Kroeber's biography proceeds with several chapters that supply in unflinching detail the decades of killing and death through the spread of disease, enslavement, prostitution, rape, and forced migration that led to the destruction of the Yahi People. Ishi's Yahi community formed a subgroup of the Yana People, who occupied an area of about 2,400 square miles at the foot of the Sierra Nevada Mountain range. Theodora Kroeber estimates the Yana People numbered two to three thousand prior to contact with whites (43). The Yahi lived in geographically isolated areas of ribbed foothills, canyons, and rivers near the base of the active volcano Mt. Lassen in northeast California. They were speakers of a dialect of Yana that had diverged more than a thousand years ago from neighboring groups. The Yana Peoples were decimated first by disease spread after contact with Europeans. Then, in 1848, gold was found near Yahi lands, and violent settler persecution against them intensified. As the settler population grew, skirmishes with the Indigenous peoples in the area soon were supported by an active government policy of eradication of resisters and resettling those who remained onto state sanctioned reserves.

In Benjamin Madley's detailed study of genocidal acts committed against Indigenous peoples in California, he states: "Between 1846 and 1870, California's Native American population plunged from perhaps 150,000 to 30,000. By 1880, census takers recorded just 16,277 California Indians. Diseases, dislocation, and starvation were important causes of these many deaths. However, abduction, *de jure* and *de facto* unfree labor, mass death in forced confinement on reservations, homicides, battles, and massacres also took thousands of lives and hindered reproduction."[30]

Only about 1,000 Yana and perhaps 100 to 200 Yahi remained by the 1850s, when a series of armed conflicts further decimated the community. In 1853, a vigilante raid by white farmers killed at least twenty-five Yahi, supposedly as retribution for stealing cattle. The Yahi group was murdered by local civilians without any legal repercussions, and not one participant was ever prosecuted. Similar attacks on other Yana peoples continued, and in 1863, the U.S. military arrived to "remove and protect,"[31] compelling 461 Yana to resettle on foot to a reservation 120 miles away. In 1864, another group of settlers organized with the intent to wipe out the Yana. As they roamed the hills, the vigilantes would kill individual Yana people when found, and massacred large groups, including 300 who had gathered for a harvest festival (77).

Ishi is estimated to have been born in 1862. The remaining Yahi lived in a part of California that was rocky, hilly, and thick with brush, not usable for grazing or farming. In 1859, a few years prior to Ishi's birth, a small group of white settlers farming and ranching in the Yahi's ancestral lands hired guards to rid the hills of remaining Indigenous peoples. The guards came upon a Yahi village and, in a surprise attack, killed forty persons, taking scalps. In 1865, another group of Yahi were killed, and in 1867, four armed guards tracked a group of Yahi to a cave in the hills. Trapped and unable to flee, thirty-three were murdered and scalped. After the killings in the 1860s, fewer than a dozen Yahi remained. This group was seen briefly in 1870 and was not encountered by whites again until 1908, when a few surveyors happened upon their dwellings.

For most of Ishi's teenage life and adulthood, there were, perhaps, only four other surviving members of his people, including Ishi's mother and sister (or perhaps cousin) and another older man. When the surveyors, who were sent by a company interested in establishing a hydroelectric dam in the area, stumbled into the camp, everyone in the remaining group ran except Ishi's elderly mother, who could not move and lay under "a pile of skins and rags."[32] The surveyors left her alone but stole every object they could gather, including items the band depended on for their survival, such as a fire drill, a bow, arrows, a spear, moccasins, and baskets. Ishi's mother would die a few days after this encounter. When asked about this event, Ishi was extremely reluctant to recount any details, though he revealed that he never saw his sister and the other man again. By the summer of 1911, Ishi likely had lived alone for three years. It is not clear why he came to Oroville, a town forty miles south of Yahi territory.

Alfred Kroeber had been studying and documenting Indigenous peoples of California over the previous decade, having arrived at UC Berkeley in 1901 as the university's first anthropologist. Upon hearing of the Indigenous man sitting in an Oroville prison, Kroeber sent his assistant Thomas Waterman, a graduate student at the university, to Oroville to consider taking charge of the welfare of the lone man. Waterman sent back word that Ishi was "undoubtedly wild" (7). Once Waterman confirmed the Indigenous man was not only likely the last Yahi but also seemed to be willing to be "a splendid informant, especially for phonetics, for he speaks very clearly" (7), Kroeber agreed to be Ishi's custodian.

Kroeber also was in charge of curating the artefacts of Indigenous peoples in the Americas assembled by the wealthy heiress Phoebe Hearst, who established a museum in San Francisco to preserve and display her collection. Hearst had sponsored archeological expeditions at the turn of the century to sites in Florida, Peru, and Egypt to gather artifacts. She also was an active buyer of antiquities and archeological objects from dealers across the world (the museum now holds over 3.8 million objects and is said to be the largest anthropological collection in the western United States).

Hearst wanted the collection, which included objects as well as human remains, to be arranged for public exhibition and serve as a repository for academic study. These materials were amassed in support of the growing field of academic anthropology that sought to "salvage" cultural works and document languages from ancient cultures and existing Indigenous communities that were perceived to be extinct or dissolving in modernity. With these collecting practices, Pauline Wakeham remarks, "anthropologists legitimated grave robbing and the expropriation of Aboriginal cultural belongings in the name of salvaging remnants of a vanishing race. The rubric of 'rescue' paradoxically constituted yet another way to inscribe, to spectacularize, and to materially hasten the death of the conflated figures of nature and natives."[33]

The Phoebe A. Hearst Museum had been preparing to open to the public in the fall of 1911, just before Ishi came to Oroville. Upon his arrival in San Francisco, Kroeber arranged for Ishi to live in the museum and work there as a janitor, sleeping in the janitor's quarters. Although Ishi could be considered a ward of the state, he told reporters and Kroeber that he preferred to live in the museum. Ishi would spend the rest of his life in the museum, willingly participating in demonstrations of arrowhead making and other Yahi cultural activities on weekends, his days off work.

Almost immediately upon Ishi's arrival, Waterman and Kroeber sought to record and document Ishi's language and culture. Perhaps with a mixture of willingness and being without any other options, Ishi participated and aided in the anthropologists' work. Ishi allowed himself to be photographed and agreed to sing several songs, which were recorded on wax cylinders. By all accounts, Ishi appeared happy and accepting of his role at the museum and his life in San Francisco.

Alfred Kroeber insisted on paying Ishi as a worker to avoid the appearance of commoditizing his demonstrations of his ancestral cultural practices while in the museum. Kroeber already had experience with the custodial violence practiced by anthropologists. As a graduate student at Columbia University in 1897 under Franz Boas, Kroeber was tasked with the supervision of six Inuit who had been gathered by the explorer Robert Peary upon Boas's request and were to be housed at the American Museum of Natural History in New York City.[34] Within months, four died of tuberculosis. Alfred Kroeber was determined to not make Ishi's living in the anthropology museum a money-making scheme or a spectacle. However, as Cherokee scholar Jace Weaver remarks, anthropology museums at the turn of the century fostered the "era of living fossils,"[35] and the lines between Native subject and object of curiosity were continually blurred with Ishi's presence as a draw to the museum.

Theodora Kroeber devotes significant passages detailing Ishi's everyday routine at the museum. Kroeber writes that Ishi "spent a couple of hours a day on exhibition days cleaning up the litter left by visitors, particularly the classes of school children. After a few days' practice, he bustled about in the early morning, handling broom, duster, and mop like an old hand, using great care towards cases and specimens. He was good with his hands, and there was about this job as about everything he did what [Alfred] Kroeber calls a 'willing gentleness'" (142). Willing or not, the image of Ishi pushing a broom in a museum full of objects and skeletons "salvaged," bought, or stolen from Indigenous peoples and ancient civilizations across the world is haunting. Theodora Kroeber knowingly stages her prose as part of the genre of "last" human writings and documentations. These sentences detailing Ishi's custodial routines are both banal and apocalyptic, congenial and devastating, and the simple grammar belies the shattering image of Ishi "good with his hands" dusting off the spoils and ruins. It is heartbreaking to catch Ishi in his daily custodial rounds, cast as an efficient worker calmly sweeping up among the hauls of American imperialism.

Ishi's own response to living through an extinction event is ambivalent in surprising ways. Ishi is well aware of the rush to observe and preserve everything about this "last wild Indian" but also is aware of the oblivion and radical loss to which he bears witness. He reacts to anthropologists and white visitors to the museum with friendship and a willingness to demonstrate his skills. On Sundays, he would greet visitors and demonstrate stringing a bow, using the fire drill, and chipping an arrowhead, which he would gift to a viewer. On these occasions, Ishi participates in his own remembrance. He willingly tells his stories and leaves his handcrafts to the museum. But Ishi also retains his ancestral beliefs regarding the sharing of names, and his friendship with Alfred Kroeber is not the same as a full companionship, as he never reveals his name to Kroeber. He remained unimpressed by a number of key modern inventions. He shrugged at the invention of the airplane, which awed the rest of San Francisco. He treated money diffidently; when he passed away, he had over $500 sitting in his safe, half of which was taken by the state as death taxes.

Ishi shows himself open to others and, at times, to demonstrating his Indigenous cultural traditions within the confines of the museum; but, overall, he remained uninterested in the compensations and labors of Californian urban modernity. He learned only a few hundred English words and did not feel the need to travel. Alfred Kroeber wrote in a 1912 article that Ishi "feels himself so distinct from his new world, that such a thing as deliberately imitating civilized people and making himself one of them has apparently never dawned upon him."[36] Kroeber describes him as separate, aloof, but also cheerful, and doesn't consider how his diffidence to white society might be an act of resistance conveyed by this remarkably polite last figure. Kroeber published very little academically about Ishi, and added just two paragraphs about him in his major work *Handbook of the Indians of California*, but these sentences are quite personal: "He was industrious, kindly, obliging, invariably even temper, ready of smile, and thoroughly endeared himself to all whom he came into contact. With his death the Yahi passed away."[37]

The fact that Ishi never revealed his real name to any friends or anthropologists after he was taken into their care has been interpreted by many, including Alfred Kroeber, as an act of resistance and rejection of assimilation. The name "Ishi" is an anglicization of the Yana word *i'citi*, which means "man." Vizenor describes Ishi's concealing of his name as a canny "simulation" and a self-obscuring that deflects from white colonial practices of naming. "Trickster hermeneutics is the silence of his nicknames," Vizenor writes.[38] Ishi accepted this name in the company of whites, but did not call whites "man" as well. He called whites "*saltu*" which Theodora translates as "a nonhuman, a pre-human, or a spirit."[39] She suggests it also might mean ghost. Whites may have seemed nonhuman and ghostly to Ishi, perhaps because they did not adhere to Indigenous rituals respecting the dead, or perhaps because whites left a legacy of death as they settled on Indigenous lands. Ishi turns the tables on the vanishing Indian trope; there is something empty and vanishing about whites, and they are identified as being the source of things that vanish and as living among the realm of ghosts. While Ishi agreed to live the rest of his life at the Hearst Museum, a place that also kept collections of bones of non-Western peoples and, thus, a location laden with ghosts, he refused to make his sleeping quarters in areas adjacent to human remains. He told Kroeber that the human remains needed proper burials and were contaminating the museum.

After much cajoling from Alfred Kroeber to visit his ancestral homelands in northeast California, Ishi agreed in the summer of 1914 to accompany Kroeber, Waterman, and the doctor, Saxton Pope, who worked in the hospital next to the Hearst Museum and had befriended Ishi (Pope also brought his teenage son along). Ishi knew he was on exhibit wherever he went, that the camera extended the exhibitionary space of the museum, and he did not want to revisit his home where his ancestors had died. Yet, ultimately, he was persuaded, and accompanied the group primarily as a gesture of friendship. Theodora Kroeber writes of this visit as a kind of psychoanalytic experience for both Ishi and his friends. "The value of the trip was neither geographic nor linguistic, but had to do rather with one man, Ishi, and was of psychological, not ethnological import. Going back to the old heartland in the company of three living people who meant the most to him would seem to have been an adventure emotionally akin to a psychoanalysis" (216). Following Freud's own interest in anthropology, discussed below, Kroeber already inscribes Ishi's traumas into a narrative of a therapeutic journey of psychological discovery through regression and re-enaction.

Over a hundred photographs were taken of Ishi on site showing him hunting, fishing, and involved in other daily life activities. On occasion, Ishi was asked to remove his clothing and live as he did prior to contact, but he insisted on keeping a loin-cloth (Ira Jacknis points out neither nakedness nor the loin-cloth are traditional Yahi dress[40]). During stretches of the day, his white friends went completely naked. There is one remarkable photograph made during the trip (now housed at the UC Berkeley Bancroft Library) showing Ishi clad in the loin-cloth looking at a naked Alfred Kroeber, who sits with his back to the camera, and a frontally exposed naked Waterman, who is standing. The photographer, Saxton Pope, situated behind Ishi, is observing Ishi in the act of observing the naked anthropologists.

The scene this photograph stages is rather Whitmanian, showing the full-frontal nudity of male bathers by the shore. Ishi, the always-exhibited, is watching a scene of exhibition of the white male "wild." The blatant playing of anthropologists "going Native" intertwines with the erotic display of the male body before the camera. The anthropologists are acting out fantasies of regression that idealize a homosocial bond that would traverse ethnographic distinctions. There has been important scholarship in the past decade focused on combining queer theory and Indigenous studies, suggesting that diverse Indigenous forms of sexuality and kinship do not fit the model of heteronormative coupling.[41] Kroeber's party is not interested in investigating any Indigenous forms of sexuality. Instead, they pursue their own sense of the erotic experience by traversing anthropological categories of "primitive" and "civilized" in a weekend. This image shows a display of male eroticism as supposedly the *homos*, the common ground between Ishi and the white male scholars.

In early 1915, Ishi began to show signs of a respiratory illness. The summer of that year, Kroeber asked Edward Sapir to come to Berkeley to work on recording and deciphering Ishi's language. Ishi enthusiastically participated in this research, but by early 1916, his illness became much more pronounced, and he was diagnosed with tuberculosis. He was admitted to the hospital next to the museum in San Francisco, where his friend Saxton Pope attended to his care. Ishi died in the hospital on March 25, 1916.

When Ishi became ill in late 1915, Kroeber was away in New York on sabbatical for the year. Understanding that Ishi's death was imminent, Kroeber wrote a letter to his assistant curator at the museum, E. W. Gifford, stating that he objected to an autopsy and any preservation of Ishi's remains for anthropological study, closing with the statement: "If there is any talk about the interests of science, say for me that science can go to hell. We propose to stand by our friends."[42] Kroeber wrote the letter the day before Ishi died; it arrived too late, as Pope had already performed an autopsy by the time it was delivered. Ishi's remains were cremated, but his brain had been removed. Kroeber found it sitting in a jar on his desk when he returned from sabbatical in 1917.

Neither the autopsy nor the removed brain is mentioned in Theodora Kroeber's book or in any of Alfred Kroeber's notes. The suspicion that Ishi's brain had not been cremated lingered among Indigenous activists. In the 1990s, a group of Maidu elders, neighbors of the Yahi, sought to have Ishi's remains returned for burial and inquired into the stories about Ishi's brain but were told the brain had not been removed. Finally, in 1999, Duke University anthropologist Orin Starn located a telegram sent in 1917 by Alfred Kroeber, but not kept in his archival papers, to the researcher Aleš Hrdlička at the Smithsonian Institute.[43] Kroeber had found the brain preserved in a jar in his office upon returning from sabbatical and, although he disagreed with the autopsy, he offered the brain to Hrdlička for scientific research. Hrdlička accepted the brain on behalf of the Smithsonian and had kept it with a numbered label in a vat with other brains.

The Smithsonian, embarrassed and apologetic upon realizing it had Ishi's brain after decades of stating the opposite, sought to repatriate the brain, eventually turning it over, in 1999, to the Pit River Tribe in northern California. An Indigenous-led ceremony and burial was performed for Ishi. The anthropology department at UC Berkeley also issued a formal apology for their role in the exploitation of Ishi's remains (in 2020, UC Berkeley also removed Kroeber's name from the anthropology building).

## Alfred Kroeber's Psychoanalytic Turn

Both Alfred and Theodora Kroeber sought at key moments to interpret Ishi's life by turning to debates over the convergences of anthropology and psychoanalysis. At several points in her biography, Theodora Kroeber construed Ishi's experiences as scenes of psychoanalytic passage. The "two worlds" motif also refers to two distinct psychological realms. She mentions in the prologue that, once Ishi entered Oroville, "he had also reached and crossed certain physical and psychic limits" (10). She also speculated on the "psychic strength" (98) necessary for the endurance of the Yahi People, who survived in harsh terrain. Kroeber marveled at the mental "temperament and build" (98) to thrive in these conditions. One source of this psychological fortitude could be found in the observation that the Yahi "were not guilt ridden, nor were they 'alienated from their culture'" (98). In a sweeping gesture, Kroeber characterized all Indigenous Californians as "introverted, reserved, contemplative, and philosophical" (23). Such psychologizing borrows heavily from the stereotype of the stoic Native. She positioned Yahi consciousness as contented until tragic, in contrast to the insatiable and alienated Western mind.

Kroeber applies psychoanalytic ideas in one direction, borrowing from the Freudian milieu of 1950s America, yet Ishi's story also pushes back against how the history of psychoanalysis in several key moments posited that the past, present, and future of the human mind lay in its regressive reiteration of its "primitive" and "savage" inheritances. The imbrication of anthropological history and versions of the origin and end of the human psyche appear regularly in Freud's own writings. In his central essay "The Unconscious," first published in 1915, Freud announced, "The psychoanalytic assumption of unconscious mental activity appears to us . . . a further development of that primitive animism which caused our own consciousness to be reflected in all around us."[44] Such "primitive animism" is understood as one of the primary manifestations of what Freud called the "instinctual life" (131) of the unconscious. While Freud claimed that "the content of the Unconscious may be compared with a primitive population in the mental kingdom" (141), associating the regular functioning of the psyche in terms borrowed from anthropological discourse of primitivity, he also came to view the damaging of the psyche in similar terms.

Freud theorized trauma under different metapsychological rubrics on a number of occasions in his career, and after the devastation of World War I, his thoughts on trauma take on an increased focus in his work. Freud's psychoanalysis of trauma culminates in *Beyond the Pleasure Principle* (first published in 1920), where he reflects on what he identifies as the intertwining of life and death drives in humans and, indeed, in all organisms. While Darwin showed that speciation and extinction are connected processes involved in both the origin and end of species evolution, Freud found that both phenomena become transposed within the psychobiology of every individual organism. In rethinking the ubiquity of human traumas and war making, Freud came to observe that the erecting and sustaining of the psyche could not be separated from processes that destroy the psyche. Freud found that the psyche can feed off and achieve pleasure in acts of self-destruction, from the compulsive repetition of psychically damaging thoughts to the notion that there is an implacable discontent within civilization that is the manifestation of an anti-human tendency within the human.

Certainly, much depends on what transposing or translating a biological or anthropological condition into a psychic condition means. Although Frank J. Sulloway finds Freud to be a "biologist of the mind,"[45] Freud is careful in drafting his theories of the psyche as not quite identical with and not quite subordinate to the fields of biology and anthropology. The processes of the psyche have their own genealogy and temporality, even as Freud repeatedly references and cites biologists and anthropologists when he finds connections and inspirations in their work to his own. But, then, what is to be made of how Freud's multiple theories on trauma rely on uncovering anthropological "primitive stages" of life? In his "Thoughts for the Times on War and Death," a discussion of World War I, Freud states that trauma works like all pathologies in that "the essence of mental disease lies in a return to earlier states of affective life and functioning."[46] Freud claimed that the traumas of war and thoughts of death could be explained by the psychoanalytic understanding of regression to earlier mental states. War and death call upon the vestiges of the primitive mind that lay permanently lodged in the unconscious. "The primitive stages can always be re-established; the primitive mind is, in the fullest sense of the word, imperishable" (286). Primitivity, then, both is and is not a temporal and historical term. The primitive mind erupts in specific historical instances but is itself "imperishable" and not subject to temporality. Primitive mental states refer to earlier states of mind that were constructed in a historical moment characteristic of more rudimentary civilizations and that remain structurally present in all human psyches. War traumas and thoughts on death reawaken these constitutive and embedded earlier states of human and proto-human affective life and functioning. Such wartime experiences and thoughts on death, thus, bring the mind in touch with the anthropological origins of the human and the ends of the human.

To what degree, then, does Freud's use of anthropological terms also refer to Indigenous humans living in the past or the present? In *Totem and Taboo*, Freud's most exhaustive attempt to synthesize anthropology and psychoanalysis, Freud characterized what he variously called "primitive races,"[47] "savages" (1), and "prehistoric man" (1) as preoccupied in daily life by instinctual matters of sex and death, and prone to egoistic impulses of selfishness and cruelty. These primitive impulses persist in all of humanity; hence "Prehistoric man . . . is in a certain sense . . . still our contemporary" (1). For Freud, the psychoanalysis of individual persons is always also a psychoanalysis of the human anthropological trajectory. Freud's developmental theory of the mind, his genealogy of primitive humanity, and his claim that neuroses uncover primitive mental states are all based on a genealogical view of the human that ties together visions of the beginning and the end of humanity.

Christian Kerslake notes that, "the circularity of the relation between the origin and end of the psyche was willed by psychoanalysis, despite the viciousness of the circle."[48] Freud sought to psychoanalyze the genesis of the human, which he saw as a parallel project to anthropology, for he believed that something about the very condition and emergence of the human requires a psychoanalysis and becomes a kind of psychoanalytic scene for the discipline of psychology itself. The primal scenes of the human tie into the primal scenes of psychoanalysis. For Freud, since ontogeny recapitulates phylogeny in both biological and psychological development, the passage of the pre-human or primitive human into the modern human is the same path by which any infant develops into maturity. Freud would come to find that the primitive origins of the human were themselves traumatic and that uncovering this primitive mind could bring knowledge about the psyche but could never dispel the original traumas that contribute, on a continual basis, to both the beginnings and ends of the human.

At some point during his friendship with Ishi, Alfred Kroeber began a more dedicated study of Freudian psychoanalysis. At the time of Ishi's death in March 1916, Kroeber was travelling through Europe, remarkable during World War I, and then in New York City, where he, ultimately, heard of Ishi's death. In 1917, Kroeber took a temporary position at the Museum of Natural History in New York, and during this time, he underwent a sustained period of psychoanalysis with Dr. Smith Ely Jelliffe, an early advocate of Freudian methods in the United States. Shortly after returning to Berkeley, and after a year of reading a number of Freud's works and other psychoanalysts, including Carl Jung, A. A. Brill, and Ernest Jones, while embarking on an analysis of his own dreams, Kroeber took full leave of his academic position and opened up, in 1920, what is thought to be the first psychoanalytic clinic on the U.S. West Coast. He spent nearly three years as a practicing analyst, serving patients who were sent to him primarily by Saxton Pope. Kroeber also became a clinician, in part to continue his own self-analysis, writing to Sapir in 1920, "I don't know whether I'm neurotic. Two psychiatrically minded doctors who've had me under treatment and know me pretty intimately, say I'm not. But I have always normally been in some internal conflict."[49]

During this period, Kroeber completed his major work *Anthropology*,[50] a comprehensive study of the field aimed primarily at a university student audience. In 1923, Kroeber returned to his academic position and would go on to several prolific decades of anthropological work. There is no direct evidence that Ishi played a significant causal role in Kroeber's undergoing analysis and becoming an analyst himself. But there is something in Kroeber's experience with Ishi that helped precipitate interest in what psychoanalysis had to say about anthropology and the primitive mind.

In 1920, Kroeber wrote a short review essay on Freud's *Totem and Taboo*, which was widely read in both anthropological and psychoanalytic circles and piqued the interest of Freud himself. *Totem and Taboo*, published in 1913 and first translated into English in 1918, comprises four chapters, but it is the last chapter upon which Freud's entire argument turns. In the initial chapters, Freud centralized totemism—defined as the belief that a community has a sacred and filial bond to a particular animal or plant—as the core organizing unconscious principle of primitive life. Freud deciphers totemism as akin to dreamwork, seeing in the totem the expressions and regulations of sexual life and the taboo against incest within the clan. Freud is aware already that there are disagreements among anthropologists as to the ubiquity of totemic beliefs across Indigenous cultures and that there is no interpretive consensus on the social functions of totems. Freud is convinced that totems as well as taboos have their origins not in the conscious realm of religions or morals but "have

no grounds and are of unknown origin."[51] Their real origins are unconscious (31) and the ambivalent compulsion to follow such totems and taboos and also the desire to break them "must be the oldest and most
powerful of human desires" (32). Taboos also are at the origin of the con-
science, the rejection of an internal wish and the simultaneous feeling of
guilt for desiring fulfillment of the wish. Freud admits there is no precise
historical or empirical evidence for these anthropological–psychological
origins; they are reasoned by inferences in assessing contemporary anthropological research and the clinical study of obsessional neuroses in
patients.

The whole of Freud's argument presumes the doctrine of psychic unity
across all humans, past and present. In the last few pages of the book,
Freud sought to offer a radical new genealogy of the human. He begins
by borrowing from Darwin the notion that since "higher apes" live in a
"horde" (125), there must have been a similar horde of primitive humans.
Within this horde, the strongest male had sexual priority over the females
and blocked the libido of the other males. Freud hypothesized that the
male members or "brothers" of the horde decided to band together to pool
their strength and intelligence to dethrone the powerful patriarch. Together, they killed the patriarch and consumed him, thus each identifying with the patriarch. To avoid a new patriarch arising, they each
renounced sexual relations within the horde, thus instituting the incest
taboo. But the brothers also began to feel remorse over the killing of the
primal father. These primal brothers, then, "thus created out of their filial sense of guilt the two fundamental taboos of totemism, which for that
reason correspond with the two repressed wishes of the Oedipus complex" (143). This guilt formed the basis of the establishment of totemism
as a social and psychological framework in which a clan worshipped a totem animal as a substitute for the father. Freud noted that, in anthropological studies in totemism, one is forbidden to kill the totem animal, yet
also there are cases that the killing of the totem is allowed on special occasions. The ensuing sense of guilt for committing murder on the father
is displaced onto an idealization of the totem and forms the basis for the
origin of moral behavior. For Freud, totemism is the earliest form of
religion, but the same core principles can be found in monotheism.

Kroeber, deeply invested in psychoanalysis personally and profession-
ally by 1920, felt compelled to write a brief but powerful refutation of *Totem and Taboo* as an anthropologist. In his essay "Totem and Taboo: An
Ethnologic Psychoanalysis" (published in *American Anthropologist*), he
called the book an item of "speculative anthropology"[52] based on conjectures and hypothetical premises posited by other theorists of human

origins that amplified these "fractional certainties" (303) into ethnological generalizations. Kroeber lists eleven points of criticism of these conjectures. Concerning the primal horde, Kroeber asks if Freud is right that guilt would be enough to stem sexual desire for members within the clan. Kroeber also points out that if the brothers did renounce women in the clan, they would all have to marry out of the clan and, thus, retain no sense of solidarity as brothers. Many of these points stress that even if Freud was right about one of his conjectures, by piling conjecture on top of conjecture, the overall likelihood of each being right in succession decreases. But Kroeber did agree with Freud's clinical insights into the "parallel" (304) aspects of the ambivalence of taboos and the behavior of neurotics. In particular, Kroeber noted that wherever the taboo on speaking of the dead "is in force" (304), it is likely that ambivalent unconscious formations are at work that include "manifestations of hostility to the dead" (304). Kroeber may have been partly influenced in this position by Ishi's insistence on not speaking of his deceased family and his reluctance to return to his ancestral lands. Kroeber concluded his essay by applauding the convergence of psychoanalysis and anthropology, but he insisted that the work of reconstructing the origins of psychic and social phenomena needed to be based on anthropological and ethnographic data.

Freud either read or heard about Kroeber's essay and provided a brief response in *Group Psychology and the Analysis of the Ego* (published in 1921), where he says of his *Totem* book, "To be sure, this is only a hypothesis, like so many others with which archeologists endeavor to lighten the darkness of prehistoric times—a 'Just-So Story,' as it was amusingly called by a not unkind English critic; but I think it is creditable to such a hypothesis if it proves able to bring coherence and understanding to more and more new regions."[53]

Kroeber's preference for historical anthropology over psychoanalysis is predictable, while Freud intentionally put at the heart of psychoanalysis an origin story that could not be verified by any empirical anthropological evidence. In *Totem and Taboo*, Freud moved between psychological historicism and metapsychological ahistoricism, empirical study and myth, anthropological data and the "translations" of the unconscious. Freud methodologically insisted on the refusal of the reduction of these dualisms. The psychoanalytic method is the disjunctive synthesis of these disparate "original" sources. There is no way to empirically verify his genealogical story, yet in Freud's account, the story has direct empirical effects on the formation of the psyche even today. Derrida, in *Archive Fever*, calls this an "impossible archaeology"—"this painful desire for a return to the authentic and singular origin, and for a return concerned to account for the desire to return."[54]

In Freud's account, becoming human had to occur through a series of renunciations using defense mechanisms that must have already been available to the proto-human psyche. The repression of sex and aggression allowed the full human psyche to emerge, yet the capacity to repress at all already presumed a human psyche capable of self-restriction and self-conflict. How would the primal horde know that the full range of the human psyche would emerge from the primal horde's killing of father and from the incest taboo? How did these human figures anticipate their own humanization? How did primitive humanity know that killing a human would allow for the civilized human to emerge? It should also be pointed out that this is strictly a male fantasy—while the brothers are bonding and killing the father, what are the women doing? These are unresolvable enigmas, perhaps "painful desires" that draw the inquiry into the psyche both closer to and further from the origins of the human at the same time.

Freud continued to maintain this hypothesis of the primal horde throughout his career, but shortly after *Totem and Taboo*, he began to combine theories on the origin of the human with speculations on the end of the human. Freud's thoughts on death culminate in *Beyond the Pleasure Principle*, which announced a revision of his previous metapsychology theory of the single libidinal drive (the pleasure principle), now replaced with a theory of the dual drives of the life drive and death drive (later recast as Eros and Thanatos). This work extends the methodological combination of myth and biology as irreducible to each other in processes of the emergence as well as the quiescence of the psyche. Freud pursues a remarkable elaboration of the possible biological genesis of this other "death drive" that exists before and independently of the libidinal drives toward pleasurable wish fulfillment. However, Freud also proposed that both drives shared a similar circuitous tendency or "urge"[55] to restore an earlier state—specifically to return the organic to an inorganic state and to quell the excitations of the drives themselves. The progress of the drives loops through their regression. If so, the very first life drives were, paradoxically, also death drives. The origin of the psyche is also coincident with the ends of the psyche. To think about the emergence of the mind is bound up with thinking about its pathways toward extinction. The death drive, which is a transfer point between life and death, points the way to self-knowledge and self-undoing.

Freud is quite animated and detailed at the prospect of citing a biological basis for these drives in the work of early genetic researchers, especially August Weismann's investigations. Yet his brief citation of

Weismann's work proves only to trouble the terms of life and death beyond their organic definitions. Weismann indicated that the "germ plasm" (later identified as DNA) could replicate itself indefinitely. If the rudimentary nucleic material appeared to be "potentially immortal" (83), then death arose as a phenomenon only among multicellular life. The original state that the drives may strive for, then, would not be death but this state of "immortal" repetition that is neither properly alive nor dead.

"Accordingly, if we do not desire to abandon our hypothesis of death drives, we must associate them from the beginning with life drives. But it must be admitted that in that case we are working on an equation with two unknowns" (93). Freud admits at this point that he is not sure what life and death mean in a schema in which they are not polar opposites but coexist paradoxically. It needs to be emphasized that Freud did not hold out for any definitive answers to the drives to be proven from genetic science and he remained unconvinced that psychoanalysis would need such a strictly empirical basis for the origin of the psyche. Psychoanalysis combines the scientific and the figurative, pairing physical stimulus and interpretation, and Freud stated that "one cannot pursue this idea [of the drives] without repeatedly combining the factual with the speculative" (95). However, Freud does mention that "biology is truly a realm of unlimited possibilities" (95), and he does not object to biological research eventually proving his theory mistaken. Refusing biological reductionism, Freud still seeks a biologically-informed knowledge that contemporary readers have been interested in transforming into a model of the psyche shaped by ecological vicissitudes.

This methodological difference also means the death drive should not be confused with biological death (and the life drive is not biological life). Paul Ricoeur calls Freud's method "metabiological"[56] and adds that Freud is consistent in his claims that there is no direct access to the causal forces of the unconscious, only indirect access via the mediations of analysis and translation (borrowing from Ricoeur, *Totem and Taboo* can be called meta-anthropological). Freud cites Darwin at several junctures throughout his career, but *Beyond the Pleasure Principle* does not discuss Darwin's own analysis of extinction, nor does Freud even mention the word *extinction* in the context of the psychoanalysis of life and death drives. The death drive harkens an individual and a species backward toward more primal forms, a sort of phylogenesis in reverse. This is as much a theory of atavism and regression—in origins that converge with ends— as it is about the phase shift from the organic to the inorganic that demarcates death.

Freud distinctly refrains from any sustained analysis of the biological science of extinction. But this is par for the course, as the death drive does not directly refer to biological death or species extinction. More recently, however, readers of Freud have sought to understand his conceptualization of the death drive as a theory of species death. For example, Ray Brassier folds Freud's theory of the death drive into Brassier's own philosophical argument that extinction entails the ultimate nullity of the philosophical inquiry into the meaning of being. According to Brassier, Freud acknowledges the "realist thesis"[57] that the inorganic precedes the organic. Brassier extrapolates from Freud's death drive this consequence: "Death, understood as the principle of decontraction driving the contractions of organic life is not a past or future state towards which life tends, but rather the originary purposelessness which compels all purposefulness, whether organic or psychological" (236). But Freud is not a strict scientific realist, and his speculative metapsychology actually moves him to problematize any clear distinctions between purpose and purposelessness, life and death, or origins and ends. Brassier may be right that extinction is a "transcendental trauma" (234), but Freud also remained convinced of another transcendental, what he called "eternal Eros."[58]

Even at a formal and stylistic level, Freud's analysis of the death drive does not cast its lot with ultimate physical annihilation, and the text itself is written in a self-critical, studious prose that recommends "patience."[59] Freud eschews the expected rhetoric of apocalyptic ends in his discussion of how the death drive undoes us all.[60] Appropriately, the text is as circuitous as the drives. He also builds his own skepticism into his arguments: "I do not know how far I believe in them" (95). One can partly explain this equanimity in Freud's conclusion that the life and death drives always are acting simultaneously. Yet the death drive, like the inevitability of species extinction, is yet another blow to the narcissism of the human. Humans do not have dominion over their own psyche or even over their own mortal condition. The circuitous death drive thwarts the aim for mastery over oneself and the world in life and in death.

Lack of mastery forces the self to care for the self and for the world, and the work of care (including therapeutic analysis) cannot be reduced to individual control. Freud would go on to connect the death drive to the "discontent" in civilization, the aggressive and antisocial behaviors that disrupt the equally potent tendency toward peaceful unity and collective achievement. The negativity and hostility that are hallmarks of the death drive can play themselves out in experiences that can be self-destructive but also self-transformative in allowing libidinal conflicts a wider range of outlets. What the death drive does direct us to do is to compulsively repeat the traumatic force of its own constitution, a trauma that makes and unmakes us at the same time. The drives beckon the psyche and the organism toward its own origins and ends, but also disrupt what we understand by these very terms. Instead of fixed beginnings and final conclusions, the drives work recursively on themselves in a way that makes it possible to negotiate these origins and ends into lived experiences of these finitudes, which are inconclusive until the very end.

## Toward an Anti-Colonial Psychoanalysis and Anthropology

Freud's death drive is as much a "phylogenetic fantasy"[61] of human ends as his theory of human origins. This metapsychological theory of the drives does not provide an unwavering script for extinction or fatalistic regression to the inorganic. Freud is clear that his metapsychology remains dualist and that the drives themselves, if not the organic constitution of life, are, perhaps, "immortal" in that they are tendencies implicit in any being responsive to stimulus as such and are neither properly biologically alive nor dead. The life and death drives draw from but also trouble biological definitions of life and death (in ways similar to "primitive" animism)—a troubling that extends to conceptualizations of extinction as well as to notions of psychical firstness and lastness. Both mortality and immortality, as well as the pleasurable drives and the death drives, are best understood as psychoanalytic phenomena in a similar way that existential thought treats biological phenomena not as fixed determinations but as interpretable "conditions." Thoughts of everything being dead (we are mere material; in effect already extinct) prove to be as uncanny as the thought that everything is alive. As such, the psychoanalytic understanding of species finitude forms yet another response to the question "What is extinction?" that requires continual reinterpretation.

Freud essentially has two theories of trauma: 1) trauma as a psychic wound or breech caused by a specific event and endured at the individual or collective level; and 2) trauma as something ontological and inherent in the organic and psychic condition that is structural to individual and collective psychology. Freud hoped that, by recognizing the traumatic, atavistic, and aggressive tendencies in technologically developed civilizations, his work would help reveal a common psychic condition shared among diverse human cultures. Finitude could, then, be incorporated in everyday self-analysis, rather than projected onto others. Trauma, the death drive, and the psyche's obsessive and contested relationship to its own extinction would be recognized as the ba-

INDIGENEITY AND ANTHROPOLOGY IN LAST WORLDS / 101

sis for commonalities across all cultures. This would go some way in deflating the pretensions of superiority of white Western supremacist declarations. However, Freud's own theories of the primitive mind are laden with colonial hierarchies and Eurocentric views of indigeneity. Although Freud pursued a kind of reawakening of the "primitive mind" in the charge of psychoanalysis, he evinced no real commitment to comprehending the diversity and specificity of Indigenous cultural practices. Freud never sought to disentangle his concepts of psychic regression, atavism, and primitiveness from associations with colonial and eugenic categories of racial difference even as he believed in psychic unity across all peoples. In *Totem and Taboo*, he drew from several sources steeped in colonial rhetoric, and declared that he would base his theory of "the psychology of primitive peoples" in large part on "the tribes which have been described by anthropologists as the most backward and miserable of savages, the aborigines of Australia."[62]

Ranjana Khanna argues that Freud's psychoanalysis is mired in "a colonial discipline" such that "it establishes a form of analysis based in the age of colonialism and constitutive of concepts of the primitive against which the civilizing mission could establish itself."[63] Julia Emberley discerns in Freud's theories of totemism how "Indigenous societies become a reductive non-embodiment of a non-European Other in which to locate the traumatic effect of the violence of the European bourgeois state."[64] It is not enough, then, just to point out the self-deconstructing motifs of origins and ends in Freud's work as evidence of shared psychic troubles across humanity. Freud's phylogenetic theory and its reliance on anthropological discourses of primitivism need to be understood in their methodological complexities and historical accountabilities.

Although Freud's own psychoanalysis of conjoined primitivizing and civilizing impulses is not the same as European imperialist territorial expansions based on violent primitivist and civilizing agendas, his theories drew from Victorian-era biopolitical doctrines that bolstered colonialism and perpetuated extinctionary practices. Freud's trauma theories are so metapsychologically and phylogenetically speculative that they miss any substantial critical engagement with the psychic effects of five centuries of colonialism and the ongoing subjugation of Indigenous peoples who are said to be furnishing the evidence for Freud's psychic paradigms.[65] Freud's theories of trauma are themselves embedded in traumatizing discourses, including the colonial practice of projecting finitude or fatality onto marginalized and racialized others—a misuse of the "death drive" concept but also an expression of that drive's anti-human aggression.

Addressing the colonialist history of psychoanalysis coincides with asking questions about how to resituate discourses of psychoanalysis and extinction in historically specific situations of trauma spurred by doctrines of firsting and lasting. Leilani Salvo Crane, a mixed-raced psychologist, poses the question of the legacy of psychoanalysis and race thus: "How do we admit the harms and complicity with the systems of oppression that have been wrought by psychoanalysis without centering whiteness? How do we create an inclusive psychoanalysis that does not marginalize, harm, or destroy non-white, non-heteronormative, gender-nonconforming populations?"[66] Building on these questions requires asking: What would a psychoanalytic theory of the origins and ends of the psyche be like if it grappled with the centuries of colonialism, the relentless and genocidal destruction of Indigenous peoples, and racist perpetuation of the division of human communities into primitive versus civilized? What would psychoanalysis be if it constructed its core metapsychological concepts of aggression, trauma, and death around long-standing and ongoing racial supremacist violence and extinctionary agendas? Could there be an anti-colonial psychoanalysis that would emerge out of a simultaneous critique and extension of Freud's attention to finitude, otherness, and non-Western lineages? How can psychoanalysis reject the colonialist discourse of firsts and lasts while still seeking to investigate extreme psychic states of origins and ends, including the effects on the psyche of thinking extinction?

I have taken this long detour through the Freudian psychoanalysis of origins and extinctions to elucidate how anthropological investments in indigeneity, wildness, firstness, and lastness resonated with psychoanalytical theories of trauma and psychic ends—and these conjoined theories have themselves produced traumatic legacies in their participation and interpretations of historical extinctionary events. "Salvage" anthropology and psychoanalysis drew attention to exploratory conceptions of the origins and ends of the human. Yet these two disciplines have themselves at times exacerbated colonialism by not fundamentally analyzing their own complicities with the oppression of Indigenous peoples by affixing upon them the terms of finitude.

Alfred and Theodora Kroeber's interests in connecting anthropology to psychoanalysis, while prompting a reflection on Ishi's own self-characterization and patience, did begin the work to raise questions about the methodological responsibilities of both disciplines, but their writings lacked a fuller critique of the implications of their own work in colonialism. In 1925, Alfred Kroeber wrote in his *Handbook of the Indians of California* that he had determined the Ohlone (Costanoan) Peoples dwelling in the San Francisco Bay Area "extinct so far as all practical purposes are concerned."[67] Kroeber did not detail the genocidal practices that led to the death and displacement of Ohlone communities. Kroeber's authoritative judgment on both indigeneity and extinction contributed to the Ohlone losing federal government recognition. Although Kroeber later revised his assessment of the Ohlone in the 1950s (and some Ohlone communities did regain status), his initial declaration had solidified the dispossession of unceded lands.

Theodora Kroeber began writing her book a decade after World War II, and several commentators have noted that her book must be seen as part of a cultural moment changed by the traumas of genocide in Europe and more capable and willing to recognize genocide within the Americas. Yet Kroeber, for her part, only briefly alludes to such connections in her work. More recently, the UC Berkeley anthropologist Nancy Scheper-Hughes wrote that the example of Ishi's experience should have provoked more reflection on extinction events and the complicity of the anthropologist. "Modern anthropology was built up in the face of colonial and postcolonial genocides, ethnocides, mass killings, population die-outs, and other forms of mass destruction visited upon the marginalized peoples whose lives, suffering, and deaths have provided anthropologists with a livelihood. Yet, despite this history . . . anthropology has been, until very recently, quite mute on the subjects of violence and genocide."[68]

Alfred Kroeber did not address how his own work participated in agendas of preserving and displacing into academic and archival repositories the cultural practices of Indigenous peoples. As a source for self-critique and self-analysis, Kroeber periodically turned to Freud. While he was critical of Freud's primitivisms—in an assessment of his own critical appreciation of Freud, Kroeber remarked that "to believe in the historically oriented twentieth century in 'the origin' of anything is a sort of infantilism"[69]—he hoped for a convergence of the two fields. In his comprehensive work *Anthropology*, Kroeber sprinkled a number of Freudian observations throughout but also stated that "the relations of anthropology and psychology are not easy to deal with. Psychologists began by taking their own culture for granted, as if it were uniform and universal, and then studying psychic behavior within it."[70] One section of the book, "Problem of the Death of Cultures," has a particular resonance for this chapter. Kroeber pushes back against the crudeness of Darwinian theories postulating as evolutionary law the extinction of cultures whereby the more technologically advanced eradicate other civilizations. Kroeber indicates that, in many cases, the "death" of a culture should be regarded as a metaphor, and he argues that "cultural content" (382) belonging to a disrupted group or nation can follow multiple pathways, including being "changed, replaced, and lost" (382). Some of these contents do "dissolve" while other "elements of the content of such cultures may have previously spread to other cultures and survive there" (382). Kroeber points to other forms of ethnic and cultural transmission and modification through diffusions "in ever new settings and with endless remodelings, selections, and recombinations" (384).

Applying Kroeber's own ideas to Ishi's case would mean shifting away from models of terminality and metaphors of lastness to understanding his story as connected to continuing efforts of cultural retransmission and "ways of going on" (384), to use Kroeber's phrasing. Kroeber concludes the section by wondering if there will be an end to all of humanity. In his brief thoughts of human extinction, he admits of multiple precedents for the dispersal and dissolution of civilizations, yet wonders if such patterns might change in the future with "a new influence of culture-planning" (384) that would create a unified, single, planetary culture, until some geological event eventually destroys the Earth. Kroeber concludes, somewhat in Freudian fashion, that he is neither optimistic nor pessimistic about such speculations, "But the question is worth thinking about" (385).

Recognizing the acts of genocide that led to Ishi being the last Yahi requires careful attention to extinctionary practices that are historical and political and that have contributed to narratives that seek to make sense of extinction, including those developed by anthropologists and psychoanalysts to explain the demise of a people or a culture. Such understanding must resist mythologizing the acts that lead to genocide while also recognizing that the "logic" of extinction has been directed by mythologies of racial, anthropological, psychological, and biological fatalism promoted under colonialism.

Roxanne Dunbar-Ortiz, in *An Indigenous Peoples' History of the United States*, argues that accounts of extinctionary genocide and survival need to be told in tandem: "Indigenous survival as peoples is due to centuries of resistance and storytelling passed through the generations, and . . . this survival is dynamic, not passive. Surviving genocide, by whatever means, is resistance: non-Indians must know this in order to more accurately understand the history of the United States."[71] Ishi's experience of genocide and survival includes his own response in making his traumatic history meaningful through telling his own stories and cultivating reciprocal relationships and friendships. His reply to the mythologies of lastness is to forge ongoing recognitions and survivances by developing new ways of responding, sharing, and imagining a future for Yahi culture. Instead of standing as an isolated, last figure exemplifying melancholic finitude, Ishi transformed himself toward remembrance and revivification of his Indigenous cultural ways. James Clifford, in his reflections on Ishi, comments that Ishi's "story is unfinished and proliferating."[72] One of the most compelling effects of these ongoing re-tellings, Karen Biestman observes, is that "Ishi became a catalyst for accountability and integrity."[73]

# PART II

# 3 / Literary Extinctions and the Existentiality of Reading

What does it feel like to read about one's *own* extinction? Why would people want to read, over and over again, about their end as a species? What happens to literature when readers read about a world without literature? These questions, tying fictional limits to human existential limits, surged noticeably at the end of the nineteenth century as a new literary culture around extinction began to coalesce based on a consciousness of the science of species finitude and the political and ecological realities of the spread of colonialism across the globe. The growing public interest in evolution furnished both curiosity and anguished inquiry into how and by what means extinction eventually might sweep away all species. H. G. Wells, in a 1902 lecture at the Royal Institution in London, wondered aloud how it could be that humans were not already extinct:

> It is impossible to show why certain things should not utterly destroy and end the entire human race and story, why night should not presently come down and make all our dreams and efforts vain. It is conceivable, for example, that some great unexpected mass of matter should presently rush upon us out of space, whirl sun and planets aside like dead leaves before the breeze, and collide with and utterly destroy every spark of life upon this earth. . . . It is conceivable, too, that some pestilence may presently appear, some new disease, that will destroy, not 10 or 15 or 20 percent of the earth's inhabitants as pestilences have done in the past, but 100 percent; and so end our race. . . . There may arise new animals to prey on us by land and sea, and there may come some drug or a wrecking madness into the

minds of men. And finally, there is the reasonable certainty that this sun of ours must radiate itself toward extinction. . . .

And yet one does not believe it.[1]

Wells's statement at the outset of the twentieth century highlights the gap between the extinction of the human species as "conceivable" yet not fully believed. Closing this gap, Wells sensed, required a new kind of fiction and literary imagination that could reconcile scientific conceptions of extinction and an appreciation of species finitude in everyday life. In the burgeoning genre of the "scientific romance," one of the precursors of modern science fiction, Wells found fertile ground for new plots based on freshly established biological theories of the origins and ends of life. At the same time, narrating one's own extinction required raising new questions about how conceptualizations of extinction affected literary form itself.

With the invention of what could be called the genre of "extinction romance" at the turn of the century,[2] authors of these narratives confronted the apparently paradoxical yet creatively generative situation in which fictional stories of extinction launched a new genre while at the same time scripting the end of all genres. Fictionalizations of human-wide extinction—here, my example is H. G. Wells's *The Time Machine* (published in 1895)—suddenly pressed upon author and reader the need to reckon with the tremendous and vertiginous implications of human finitude that could call into question every aspect concerning narrative form, content, and the foundations of literary genre. Wells's work presented extinction as happening in both real time and deep time, cleverly using the narratological device of time travel to consolidate these disparate temporalities. By the end of the nineteenth century, after just a few decades of scientific evidence that began to reveal the extent of the effects of extinction across the earth, a vision of the traumatic end of humans became an idea addressable in everyday encounters and in characteristic literary styles. Extinction caught hold of people's imagination but also drew minds toward the unimaginable—and the rise of science fiction arrived just in time to document and experiment with the possible/impossible question: "What is extinction?"

It becomes feasible to recognize the literary value of this question at the end of the nineteenth century for reasons internal to the formalist potentials of science fiction that began to prove themselves especially capable of mapping and making sense of planetary-scale ecological distresses. Simultaneous with the development of Victorian-era extinction narratives that incorporated such extreme narratological situations,

Mike Davis discusses how a series of "late Victorian holocausts"[3] occurred across the globe, marking the advent of a kind of global apocalyptic experience of human suffering through ecological disasters exacerbated by political oppressions. Davis documents how a series of extreme drought events caused massive global famines in 1876, again in 1889–91, and again in 1896–1902. The countries of India, China, Brazil, Ethiopia, and Russia were hit particularly hard by collapsing agricultural production. The misery wrought by drought was compounded with widespread epidemics of malaria, bubonic plague, and dysentery, culminating in the deaths of an estimated 30 to 60 million persons. The proliferation of new apocalyptic narratives at the limits of the literary in European and American letters at the turn of the century requires contextualization along with the apocalyptic history of what Davis describes as "a new dark age of colonial war, indentured labor, concentration camps, genocide, forced migration, famine and disease" (139).

Understanding the conceptual history of extinction requires tying the science of species ends and the formal and cultural effects of extinction with specific political and global social events in which decisions over life and death are continually being made. Previous chapters have examined how extinction events concerning humans and nonhuman animals intertwined with naturalist discourses and aesthetic forms in ways designed to make the extinction of "others" documentable and manageable. Wells's sense of imminent human disaster situated at home in London patterns its narratological extremes in part on the historical extremes of human suffering that ensued in the late nineteenth-century rush toward planetary-scale imperialism. The obsessional attitude at the turn of the twentieth century to record the death and disappearance of species and races converged with the profusion of speculation on the eugenic and dysgenic plots for civilization as a whole.

Wells turned to science fiction and the extinction romance as exemplary for exploring how "novel thinking" could engage with extinction at the level of literary form and content. In a similar way that Gillian Beer finds novels in the wake of Darwin pursuing evolutionary plots around descent, natural and sexual selection, and species entanglement, the increasing awareness of extinction had powerful effects on the form as well as content of literature.[4] As we have seen in earlier chapters, what could be called Darwin's second dangerous idea in *On the Origin of Species* was to intertwine definitively speciation with extinction (his first "dangerous idea," in the phrasing of Daniel Dennett, is non-teleological evolutionary development according to processes of natural selection[5]). While voyaging on the *Beagle*, Darwin observed: "Certainly, no fact in the long history of the world is so startling as the wide and repeated extermination of its inhabitants."[6] Darwin made the case that the theory of evolution had to account for the failure of some species to evolve, and that extinction was not an aberration but an expected and common occurrence for species. To understand the origin of species, one had to come to terms with the causes of their ultimate ends. But thinking about extinction also applied to the thinker, who could not but be startled by another Darwinian dangerous idea: the possibility of an era on Earth when the human species becomes an impossibility.

Darwin's ambiguously titled study *The Descent of Man* suggested he might have examined further the implications of his own argument on the long-term future of the human species. Instead, Darwin applied his theory of extinction in this work most specifically to account for the disappearance of non-Western cultures before and after Western contact. In the opening statements of the first chapter, Darwin announces: "Do the races or species of men, whichever term may be applied, encroach on and replace each other, so that some finally become extinct? We shall see that all these questions, as indeed is obvious in respect to most of them, must be answered in the affirmative, in the same manner as with the lower animals."[7]

This remarkable claim employs a substantial amount of imprecise language, perhaps to an unusual degree given Darwin's pedigree and the stakes of his remarks. Every phrase here can be questioned as to its scientific clarity and veracity. Darwin acknowledges he is unsure what term he should be using when a distinct human community no longer exists—"races or species of men, whichever term may be applied"—but his terms conflate vastly different categories. Darwin knew there was only one human species, but, here, his language indicates he is not sure he should be thinking at the level of all humans, or what he assumes is the level of inter-human conflict of races or "species of men." Using the plural prevents the designation of "races" from being isomorphic with the totality of humanity. To "encroach" is not the same as to "replace"—but, in either case, Darwin neglects to consider how human groups might also intermix, co-develop, or absorb aspects of other groups, such that there is no "finally extinct" scenario but a continual changing and amalgamation of human lives. Some human cultures and communities certainly have disappeared, but the phrase "finally extinct" has questionable application to summing up all instances of the history of the effacement of peoples since it can be the case that elements of these peoples and cultures are passed on or still persist in other ways. The analogy that extinction among humans works exactly like "the lower animals" has more qualifications and exceptions than Darwin acknowledges.

Darwin's theorizing of extinction prompts him to consider it not as applicable to humans in general but, rather, as a biopolitical criteria to divide the category of the human and identify how some human "races or species" rather than others are liable to demise. Darwin states later in the work his definitive conclusion: "When civilized nations come into contact with barbarians the struggle is short, except where a deadly climate gives its aid to the native race."[8] How scientific is the category "barbarians" and to whom does it apply? Why is "contact" here necessarily and immediately understood as violent "struggle?" Why is "climate" an extraneous factor in struggle, since in many instances of early contact with Indigenous peoples, the survival of Europeans in harsh weather conditions depended on food and knowledge from locals? Finally, there is the question of why extinction here is foremost theorized as a struggle between different human peoples rather than a human-species wide condition? Instead of putting into question the assumptions of any civilizational stability by applying the notion of extinction to all humans, Darwin's claims imply that what he calls "civilized nations" are always on one side of the extinction ledger. The winner of extinction embodies "the descent of man," what Sylvia Wynter describes as an "over-representation of Man,"[9] a totalizing figure that announces itself as the standard definition for what it means to be human based on a biological-civilizational index of purported superiority.

## Last Storyworlds

Darwin's selective application of his own theories to the conceptualizing of human extinction points to two general models of extinction in the late nineteenth century: 1) narratives of the extinction of others understood as substantially different from the narrative self, and 2) narratives of extinction inclusive of the narrative self, which are often inflated to "over-representing" all humanity as seen from the writer's own point of view. Although many scholars have flagged Wells's story *The Time Machine* as a Darwinian-influenced extinction narrative, its plot counters Darwin's assumption of the victory of "civilized nations" and, instead, transforms Darwin's "descent" into a fictive rendering of the destructive trajectory of all humans. As the character named, simply, the Time Traveller journeys forward into deep-time future and eventually becomes witness to a time without humans, he never leaves the space of what once was London. The

Time Traveller testifies to the end of British existence, the synecdoche for all human life, undone by what he interprets as malign legacies of class antagonisms and the inability of British society to collectively solve the perceived Malthusian problem of population outpacing resources.

Wells's novel is an early example of self-implicated apocalypse, such that the end of the world is not projected on an externalized "they" or displaced on a colonized other (other examples from the period include Richard Jeffries *After London* [1885], Camille Flammarion's *La Fin du Monde* [1894], and M. P. Sheil's *The Purple Cloud* [1901]). These novels place the cause of extinction either in something inherent in the human condition inclusive of themselves or in biological, geological, or cosmological events that affect everyone. The author, narrator, and even the reader all are put into the position of being last humans. By implicating everything concerning the storyworld into the narration of extinction, readers are drawn irresistibly to question how it was possible at all to read the very work they were reading. Last readers are invited to include themselves in the extinction plot as witnesses to how the end of literature becomes its own literary experience.

Wells cannily recognized that the question of extinction potentially led to questioning everything about the workings of the novel. In last human novels, we are given a series of basic literary paradoxes: How do we read a text that, technically, should have no readers? How can a character account for the end of character? How can literature narrate the end of literature? What happens to point of view when there is no more possibility for an anthropic point of view? What are the existential limits or finitudes of the literary condition? There is a peculiar pleasure in how last human novels unravel the formalist properties of narrative in a rather formal way.

Claire Colebrook contends that literary theory needs to think beyond the end of literature to consider "texts as remaining, unread, dead objects without authors or audience. . . . A literary theory would not assume that texts or letters were the work of a living body, and yet would be theoretical as well as literary in asking what sort of reading, viewing or look such texts or marks might open."[10] In the Victorian period, readers had to negotiate simultaneously a traumatic sense of literary demise (texts as "unread, dead objects") and a new formalist aesthetic to comprehend narratives of their own extinction. This vision of one's own extinction at the end of the nineteenth century began to serve in a remarkable way as a cornerstone of aesthetic education in its simultaneous construction and dissolution of the middle-class reader.

Most of the story in *The Time Machine* dwells on the first use of the contraption to visit the year 802,701. The stated aim of this journey is to find out "What might not have happened to men?" (22). The question of the future of humanity is awkwardly posed in the negative. This query is not initially meant as a phrase to describe the end of humans. Anxious at the outset with the prospect of encountering the future of humanity, the Time Traveller steels himself for encountering a potential moral crisis besetting these humans, which he immediately ponders as involving the threat of a withering masculinity. "What if cruelty had grown into a common passion? What if in this interval the race had lost its manliness, and had developed into something inhuman, unsympathetic, and overwhelmingly powerful?" (22).

Once the machine begins to whirr into action, the Time Traveler, who is seated in place at the controls, describes the visual effect of time travel in terms distinctly akin to the effect of watching the cinema. "I saw trees growing and changing like puffs of vapour, now brown, now green; they grew, spread, shivered, and passed away. I saw huge buildings rise up faint and fair, and pass like dreams" (19). Several readers have identified the layering of the time machine's mechanical and visual effects borrowed from the technological and visual repertoire of the kinetoscope and motion photography.[11] As if anticipating the invention of time-lapse cinematography,[12] this description provides the reader a way to visualize at a glimpse the epochal arc of species time together with British civilizational time passing from birth to death. This time-lapse also allows for the observation of geological epochs passing in what can be called a space-lapse, as the Time Traveller observes the Thames River shifting course. Even long-term climate change phenomena become legible in a single sentence: "I saw a richer green flow up the hill-side, and remain there without any wintry intermission" (23).

*The Time Machine*'s time travel plot coincides with an evolutionary plot that provides access to a vision of a world without humans. By the end of the novel, set some 30 million years in the future, the reader is confronted with a devastating vision of last things—humans are long gone, life on earth is ending, and the sun is expanding. The Time Traveller stands in stunned witness: "I cannot convey the sense of abominable desolation that hung over the world."[13] The Time Traveller suddenly becomes a last human figure, a figure who supplies a fictional account of the end of fiction. Part of the thrill of apocalyptic genre fiction comes from readers being able to witness how the basic narratological components of novels are put into existential crises no matter how predictable the plot. Last fictions compose a world in which readers encounter the existentiality of narratological devices, which structure the novel but also show a world where novels are no longer possible. Reading the existentiality of narrative devices entails considering how the devices are involved self-referentially in facing their own matters of functional longevity and perishability. These devices can figure in their own disappearance by shifting into a mode of "last" or non-anthropic point of view. Reflecting on Wells's novel, thus, will require re-examining some basic precepts in narratology and literary formalism to think about extinction.

Wells's determination to provide the English public with an accessible and imaginative narrative for human extinction situated at home had a prominent role in launching his career. In an 1893 essay titled "On Extinction," published in *Chambers's Journal of Popular Literature, Science, and Art*, Wells explicated extinction in terms of middle-class cultural frames of museum leisure-time, classical literary genres, and self-reflexive epiphanies stirred by tropes familiar to British Romanticism. In the first sentence, Wells declares that species finitude and the work of tragic genres are fundamentally intertwined: "The passing away of ineffective things, the entire rejection by Nature of the plans of life, is the essence of tragedy."[14] Wells then proceeds as if strolling through a natural history museum, offering passing observations on species failures such as the pterodactyl, which had the brilliance of flight but apparently left no descendants. Abruptly shifting his gaze to the present, Wells then remarks: "In the living world of to-day the same forces are at work in as in the past. It is not only the dodo that has gone; for dozens of genera and hundreds of species, this century has witnessed the writing on the wall" (170).

The genre of tragedy, fixed and fated in biology and in classical cultural structures, continues its work in the present. Wells mentions the decimation of the bison along with a number of other threatened animals and conflates their precarious lives with "the Red Indian." For Wells, these examples are reminders of the melancholy of biology and human frailty in the midst of progress and British power at the turn of the century. Reflecting on the poem "The Last Man" (published in 1826) by Thomas Hood, Wells writes that Hood "probably hit upon the most terrible thing tha[t] man can conceive as happening to man: the earth desert through a pestilence, and two men, then one man, looking extinction in the face" (172). Wells repeatedly returns to this anguished humanistic reflection on the potential end of the human in much of his fiction.

In further evidence of his obsession for extinction and its cultural implications, Wells wrote another short essay, "The Extinction of Man," which appeared in September 1894 in the popular afternoon daily *Pall Mall Gazette*. Wells begins by addressing humans at a species level, identified with the pronoun "it": "It is part of the excessive egotism of the human animal that the bare idea of its extinction seems incredible to it.

'A world without *us!*' it says, as a heady young Cephalapsis might have said it in the old Silurian sea."[15] The vocalization at the level of the human species is simultaneously intended to be read as an imagined utterance from the Cephalapsis, a bottom-feeding fished that went extinct 400 million years ago. Wells continues: "But since the Cephalapsis and the Coccosteus many a fine animal has increased and multiplied upon the earth, lorded it over land or sea without a rival, and passed at last into the night. Surely it is not so unreasonable to ask why man should be an exception to the rule" (3).

These thoughts are planted in the midst of sundry journalism on the hum-drum affairs of British life. Wells applies his knack for bringing species crises into staid parlor rooms as the essay continues by imagining what kind of beast will eventually supersede humans. He considers cephalopods in the sea as potential future rivals, or perhaps a future dominated by massive well-organized and hungry colonies of ants, or a new disease epidemic as potential downfalls of humanity. As if an advertisement for Wells's coming fiction, his appeal for humans to consider their own extinction is matched with a scenario in which fiction would provide the staging. "In the case of every other predominant animal the world has ever seen, I repeat, the hour of its complete ascendency has been the eve of its entire overthrow. But if some poor story-writing man ventures to figure this sober probability in a tale, not a reviewer in London but will tell him his theme is the utterly impossible" (3).

Both these essays revel in Victorian palpitations toward biological and cognitive degeneration and devolution, discourses that intermingled with notions of evolution and extinction at the turn of the century. Max Nordau, in *Degeneration*, his diagnosis of a *fin-de-siècle* malaise sweeping Europe, claimed "the prevalent feeling is that of imminent perdition and extinction."[16] Nordau accused contemporary art of aggravating trends of social and physical declines in health. He lumped together movements such as symbolism, impressionism, and the pre-Raphaelites as having ill aesthetic and biological effects on audiences. What stands out in Wells's brief essays and early fiction on extinction is his sense that the human species is heading toward its downfall, not because of any invading outsiders, degenerate cultural products, or the intermixing of races but because of human overconfidence and anthropocentric egotism.

In *The Time Machine*, white British imperialism, along with the genre of scientific romance itself, are swept up in biological and planetary entropy that serves as comeuppance to a culture and species too convinced of its exceptionality. Instead of pinning impending civilizational trauma on those deemed less fit, Wells imagines no outside for Britain and no alternative to collective human extinction. As Christina Alt has shown, Wells went on in his career to reassess such biological fatalism and entertain a variety of ways the species, or some humans in particular, might escape this end. Alt remarks that "extinction, in Wells's mind, went from being feared as a threat to human survival to being viewed as a phenomenon to be harnessed by human beings so that they might decide for themselves the composition of 'nature'."[17] But in his early writings on extinction, the work of extinction is pervasive, and no salvation is posited. With British fortitude, Wells indicates one must face catastrophe head on even if it cannot be avoided, and he even suggests the failure to do so only hastens the downfall of the species.

## Narrating the Natural History of the Future

In *The Time Machine*, Wells uses the device of a story within a story to allow for two characters to serve as narrators for the novel: first, an unnamed witness who meets with a group of other local bourgeois characters is called on to testify to the credence of time travel, and then the Time Traveller himself takes over, relating his story over dinner using a past-tense verbal narrative construction. The bulk of his story covers the first use of the machine and the Time Traveller's encounter with the human-animal remnants of civilization along the Thames River. The two species descending from humans that are featured in the novel, the Eloi and the Morlocks, which have garnered so many evolutionary, sociobiological, and quasi-Marxist interpretations, also become poignant figures for extinction. They embody the mythologized conflation of evolutionary visions of pre-human and post-human life. They are characters who also characterize the end of all character and culture. The Time Traveller relates that the Eloi are illiterate. Weena, the only post-human character who is named, is unable to conceive of literature, for "the bare idea of writing had never entered her head" (64). Wells, via his protagonist, suggests that the end of literature would be a barometer for the end of humans.

These last humans find themselves enveloped in a world where the work of extinction is pervasive. The Time Traveller mentions that the Eloi eat only fruit, which he first thinks may be a sign of vegetarian utopianism, but soon learns otherwise upon realizing that the consumption of animals has been all but finished due to extinction. "These people of the remote future were strict vegetarians, and while I was with them, in spite of some carnal cravings, I had to be frugivorous also." He then adds, "Indeed, I found afterwards that horses, cattle, sheep, dogs, had followed the Ichthyosaurus into extinction" (27). Readers have to guess about cause and effect: does looming planetary extinction provoke these examples of behavioral atavism in humanity, or did atavistic activities precipitate the undoing of humanity? In any case, Wells envisions post-humanity and pre-humanity converging in a simultaneous advance and regression.

The Eloi are seen by the Time Traveller as having the cognitive capacity of five-year-old children and physically appear effeminate. They speak a language, but the Time Traveller never manages to learn it, and describes their voices as "cooing" (25). In the eyes of the Time Traveller, Weena's physical and cognitive meekness also indicate a weakness of character and species—she is barely able to be seen as a person and is called "the thing" and "the creature" (42).

The Morlocks, denizens of the underground and rivals for the claim to post-human survival, only grunt. They have no names or individuality, and their personhood is submerged wholly by an animality of the most repulsive kind, as they are called "human rats" (74) and are associated with lemurs by the Time Traveller. With their chinless faces and lidless eyes, the Morlocks, "nauseatingly inhuman" (55), are consuming the Eloi—the remnants of humanity pursuing its own eradication.

Neither of these monstrous post-humans have made a culture out of the end of culture—one of the aesthetic signatures of last human novels— as they simply no longer have the capacity for culture. The Time Traveller tries out a number of alternating explanations for how humans might have split into these two subspecies. He wonders: perhaps class struggle led to a kind of biological stratification due to capitalists being pitted against laborers, with the working class wallowing below ground and the upper-class lording over them. Then again, perhaps communism eliminated social struggle and led to a general lassitude that resulted in such grotesque and desperate beings. Perhaps a pastoral utopianism might have taken hold whereby humans found themselves contented, decadent, and no longer motivated to forge themselves physically and mentally. But he also wonders if some dark plan was set upon by the Morlocks to breed the Eloi as a passive, cattle-like food source.

Chased by the hungry-eyed Morlocks, the Time Traveller finds himself entering a cavernous building he dubs "the Palace of Green Porcelain." Inside, he finds a decrepit museum of natural and cultural history, which provides a further opportunity for the Time Traveller to reflect on the remnants of humanity.

> At first glance I was reminded of a museum. The tiled floor was thick with dust, and a remarkable array of miscellaneous objects was shrouded in the same grey covering. Then I perceived, standing

strange and gaunt in the centre of the hall, what was clearly the lower part of a huge skeleton. I recognized by the oblique feet that it was some extinct creature after the fashion of the Megatherium. . . . Further in the gallery was the huge skeleton barrel of a Brontosaurus. My museum hypothesis was confirmed. . . .

Clearly we stood among the ruins of some latter-day South Kensington! Here, apparently, was the Palaeontological Section, and a very splendid array of fossils it must have been. (64–65)

Wells frequently visited the majestic building of the British Museum of Natural History in South Kensington (now called The Natural History Museum) while a student at the nearby Normal School of Science. The massive building opened in 1881 with large zoology, paleontology, and minerology exhibits and a smaller botanical collection. For a moment, I want to consider some of the aesthetic and psychological impacts of strolling through a natural history museum, with its extinct and existing animals displayed in juxtapositions that provide a sweeping view of evolutionary time similar in effect to Wells's novel.

In a fashion analogous to scientific romance novels, the natural history museum abounds in lively animal dramatizations, species tropes, and a fascination with biological forms and finitudes.[18] Several eminent naturalists in the nineteenth century who contributed to major developments in theorizing extinction, including Georges Cuvier and Richard Owen, took it upon themselves to manage the institutionalization and exhibition of the natural history collections under their supervision. The British Museum of Natural History, dubbed "the Temple of Nature" with its Romanesque ornaments and Gothic church vaulting, promoted a quiet contemplation of nature's relics. By exhibiting large collections of bones, from dinosaurs to more recently extinct animals such as the megatherium, museums of natural history became locations for the appreciation of extinction as a spectacle and a physical thing to be approached in a face-to-face encounter.

Though natural history exhibits are not cast as intentional works of art, they beckon aesthetic interpretation with their formal attributes shared by the fine arts in their visual appeal, emphasis on materiality, and curated recreations "from life." In the British Museum, display after display of animal bones evoked a sense of both haunting and dispassionate historical knowledge, compelling the viewer to handle visions of death and the charnel house[19] in conjunction with an open-minded appreciation of biological facts and awe at nature's diversity. The bones were exhibited to emphasize a sense of immediacy and closeness to an animal that may have long vanished, and the animals often were positioned as if in active movement, showing vitality and demise at the same time. Curators at the British Museum situated for comparison the eighteen-foot body of the extinct *Megatherium americanum*, the giant ground sloth, next to the bones of the living three-toed sloth (*Bradypus*), which is dog-sized.[20] Wells's mention of the megatherium is a nod to both this display and the work of Cuvier, who first reconstructed the animal and declared it extinct. The herbivore megatherium is positioned with its hands grasping a tree trunk; the viewer can visualize the animal eating but also finds the bones evocative of a lost world of giants, suggesting the animal's hunger and size made it all the more susceptible to a traumatic end.

These museums could not help but convey melancholy toward the biological archive in portraying both speciation and extinction together. The natural history museum became, in effect, the museum of extinction, helping to cultivate a broader public culture around awareness of species finitude. Within the span of a two-hour visit—about the same amount of time it takes to read Wells's novel—viewers could encounter hundreds of millions of years of evolution and species loss. These museums contributed to the affective and scientific ground of public attention to extinction, prompting viewers to mix sentiments of wonder and terror, the study of biological form with the end of form, conflating adventure and collapse, liveliness and ghostliness.

The rise in prominence of natural history museums in the Victorian and modernist periods parallels the growth of the science fiction genre. Wells employs the scene of the ruined museum housing long-lost species and defunct civilizational objects as a meta-reflective moment in a novel that is about the ruins of human civilization. In the chapter that takes place in the Palace, Wells explicitly asks the reader to compare animal extinction and literary extinction, and to use the tools of natural science and literary enjoyment as complementary in fashioning a self-reflexive perspective on extinction that would include the cold eye of science and the hot thrill of the survival narrative. Inside the Palace, the Time Traveller stumbles upon a library teeming with "decaying vestiges of books" (67). Wells, using literature to depict a scene of literary wastage, has the Time Traveller bemoan, "Had I been a literary man I might, perhaps, have moralized upon the futility of all ambition. But as it was, the thing that struck me with keenest force was the enormous waste of labour to which this sombre wilderness of rotting paper testified" (67). The fictional Time Traveller is literally nothing but "a literary man" who ironically laments the end of "literary man." Such scenes of reading the end of reading appears often in last human novels, and the theme returns again at each stop in the future the Time Traveller makes.

The Time Traveller escapes from the cannibalistic Morlocks and abandons the cloying Eloi, but rather than retreating from the experience of human extinction, he continues further into the future to witness a world without humans and the end days of the Earth. His next stop is an unspecified time billions of days later. As he speeds into the future, the Time Traveller observes the sun has grown "broader and more red" (81) and eventually no longer rises and sets. He realizes the Earth has become tidally locked, with one side permanently facing the sun. He encounters a massive crab-like creature and has the epiphany that this future Earth, with its noxious air, oceans, animals, and plants, no longer offers a home to any kind of human.

The Time Traveller again presses on, setting the machine for a time 30 million years forward. The crabs have disappeared, and only a thin film of lichen "alone testified that life was not extinct" (84). The planet is now rapidly cooling due to the sun's fading radiant heat as it passes into its red giant phase. Struck by another planetary phenomena, the shadow of an eclipse, the Time Traveller states, "A horror of this great darkness came upon me. The cold, that smote to my marrow, and the pain I felt in breathing, overcame me. I shivered, and a deadly nausea seized me" (85). This entropic vision of a cold sun is a figure for the end of life and the end of all narration. The Time Traveller is doubly haunted by the silence of the end of planetary life and the silencing of any point of view or narrative beholder. "Silent? It would be hard to convey the stillness of it. All the sounds of man, the bleating of sheep, the cries of birds, the hum of insects, the stir that makes the background of our lives—all that was over" (85). The silence and coldness both deadens and awakens the Time Traveller; he feels himself about to collapse, but the feeling of this precipitousness revives him. The Time Traveller is an inverted Robinson Crusoe character—he is not the first but last inhabitant, not the initiator of the genre of the novel but delivering its elegiac farewell. Close to fainting, the Time Traveller pulls himself back into the time machine and flips the switches for home.

## Last Narratology

In his recent study of time travel fiction, David Wittenberg claims that time travel novels function as a kind of "narratological laboratory" such that "many of the most basic theoretical questions about storytelling, and by extension about the philosophy of temporality, history, and subjectivity, are represented in the form of literal devices and plots."[21] Last human stories work in similar ways as a laboratory for observing and reflecting on the function and limits of literary form. In reading last fictions, readers become self-reflexive about the ends of their own species form at the same time they become self-reflexive about the ends of literary form. It may seem strange to foreground matters of narratology when extinction presents such a radical and urgent situation—why would readers have time to ponder the mechanics of literary construction when their species' existence is at stake? Yet it is precisely the continual shifts between conceptualizing real and fictional ends made possible by last narratives—and the realization that the apocalyptic end of story, if it is told within the storyworld, need not actually be the end of storying—that will require readers to navigate the limit questions of literary form and species form simultaneously.

Wittenberg curiously downplays the narratological effects of time travel in *The Time Machine* and declines to cite the novella as exemplifying what he considers to be the most interesting aspects of time travel's narrative inventiveness. According to Wittenberg, "The essence of time travel fiction, for the purposes of narrative-theoretical work, lies in its specific methods of constructing and juxtaposing narrative registers, layers, or lines" (26). Wittenberg claims that Wells uses literary techniques of voyaging through time to expostulate on his central concern of elaborating an "evolutionary macrologue" (35). Reducing Wells's "late utopian romance" to an expression of "uneasy realism," Wittenberg finds that *The Time Machine* is "disinterest[ed] in possible contradictions and paradoxes" (89) that mark the more creative structures of time travel narratives.

Yet *The Time Machine* certainly does radically juxtapose narrative registers: the storyworld stretches to the entropic ends of the Earth, while the entire tale is recounted within the Victorian interior parlor room space. Wells puts adjacent deep-time and Victorian present, oscillating between sentiments and sentences of familiarity and utterly devastating experiences—while retaining the serviceable architecture of the novel. The Time Traveller often shuttles between these immensely discrepant perspectives, and characterizes his double vision with a literary metaphor poised between the legible present and illegible future: "Suppose you found an inscription with sentences here and there in excellent plain English, and, interpolated therewith, others made up of words, of letters even, absolutely unknown to you? Well, on the third day of my visit, that was how the world of Eight Hundred and Two Thousand

Seven Hundred and One presented itself to me."[22] The Victorian English sentence structure and polite address are still there, but they are hardly able to function.

The arc of the story also is not a one-way extrapolation from present to future. The Time Traveller returns to tell his tale, so that its listeners will be motivated to change their own present to alter the path of the deep future. Thus, to correctly read the novel would be to overcome atavistic tendencies in British society, making it so the future depicted in the novel will never come to fruition. That is, one must read in a way such that this particular future will become unreadable. To correctly interpret what has been witnessed in the future, one must read and act in such a way as to make that future illegible and impossible. As one of the characters states to the Time Traveller of his mechanical contraption—and by extension his whole story—"But the thing's a mere paradox" (16).

A number of recent narratology theorists propose that situations of "extreme narration," or narrative at the limits of narrative, are becoming increasingly common in fiction over the last century. Narratologists have been inspired to revisit the borders of narratology especially in response to the proliferating number of novels in recent decades featuring nonhuman narrators and non-realist points of view.

Brian Richardson calls these "extreme narratives" or "unnatural narratives" because they "violate mimetic conventions and practices of realism."[23] Aligning "natural" narration with realistic mimesis harkens as far back as Aristotle, who argued that characters imitate human actions. Formalist and structuralist theories of character generally assert that characters are textual functions or devices that produce a variety of reality effects, most particularly the effect of psychological realism. Counter to this adherence to the mimetic model, Richardson's argument stresses "unnatural" kinds of characters who we do not normally encounter in everyday life or do not have the same psychological profile as the reader. Examples include novels that posit "post-anthropomorphic"[24] or post-human speakers such as animals, machines, or inanimate objects.

Richardson also points to fictional works that use narrative devices to destabilize, contravene, or undo their own basis for narrativity. Richardson mentions Samuel Beckett-like sentences—"Yesterday it was raining. Yesterday it was not raining"—as a narration that says and unsays itself, disqualifying its own means for making narrative sense.[25] While Beckett is fond of employing philosophical paradoxes and antinomy at the level of the sentence (the classical example being "I am lying"), Wells embeds the antinomy in the narrative structure itself. For Wells, and the same goes for the structure of finitude for human species, the end of the novel makes and unmakes the conditions of its experience.[26] Wells makes the structure of legibility of the novel converge with a future without readers. The species that reads this novel is the species that will not be able to read this novel, a future that the novel patiently explains to the reader.

Some narratologists welcome the attention to unusual speakers or points of view and do not see them as limit cases or extreme examples but, rather, as part of the general paradoxical strangeness that operates in all forms of narrative. Marco Caracciolo rejects the binary implications of unnatural versus natural narration to claim that novels provide a plethora of "strange narrators" that stretch our normative expectations of how characters function.[27] Marie-Laure Ryan distinguishes between narrative worlds that cannot ever be actualized but are still legible (fairy tales, for example), worlds that are pragmatically impossible (humans transformed into animals), and worlds that are logically impossible (time travel).[28] Extinction narratives, texts with no readers that can still be read, posit logically impossible readers but are still legible and coherent as storyworlds because readers consent to have their imaginations stretched and suspended in this possible/impossible paradox.

Jan Alber argues for retaining the narrative categories of natural and unnatural because we have cognitive expectations for how narration should work in everyday life. The challenge, Albers notes, is how readers come to make sense of narratives that seem to attack or undermine the mechanisms of sense making. As Alber claims, "Unnatural scenarios and events primarily concern the question of 'what it is like' to experience the transcending of physical laws, logical principles, or standard human limitations of knowledge and ability, and such experiences are restricted to the world of fiction."[29] Extinction narratives are extreme narratives that provide the reader with the apparent paradoxical experience of "what it is like" to have no more human experience; in effect, an experience that one can have and not have at the same time. Wells makes the impossible scenario of a world without us, without readers and storyworlds, legible for the reader, an example of what Alber describes as a "humanly impossible scenario" (3) that is, nonetheless, possible to narrate.[30]

There is something particularly strange or "unnatural" that happens to the narratological mechanisms of character in Wells's novel. I have already discussed how the post-human/subhuman Eloi and Morlocks demonstrate to the Time Traveller the unbecoming of character as indicative of the path of evolution/devolution of the human. The post-humans are characters who are not quite cognitively able to recognize themselves as characters. Yet even the humans in the novel are treated as not fully characterological. Many readers have noticed the evident flatness of

Wells's characters in this novel in which practically no one is granted a family name. In the opening chapter, names are whittled down to job titles or typological roles. At the opening of the novel, along with the Time Traveller, we briefly meet a slew of persons, including the Psychologist, the Provincial Mayor, a fellow named Filby, the Medical Man, the Very Young Man, the Editor, the Journalist, and the Silent Man (the names come with the article "the"). When E. M. Forster made his distinction between "round" and "flat" characters, he singled out Wells for producing a plethora of characters who "are as flat as a photograph."[31] For Forster, flat characters are reducible to types and predictable behaviors. Round characters garner more affective attachments and seem like "real people" full of complexity.

For Wells, the lack of character depth serves a narratological purpose. Whether or not it is intentionally planned by Wells, his story employs flatness of character as correspondent to the effects of witnessing Victorian everyday time juxtaposed with the time of species extinction. Wells also applies flatness of character distinctly onto the Eloi and Morlocks. The flatness of their character signifies for the Time Traveller the self-destruction of the educated mind: "I grieved to think how brief the dream of the human intellect had been. It had committed suicide" (78). The tradition of the novel is built on the assumption that the fullness of literary character is analogous to the fullness of human experience. While round characters are the staple of the *bildungsroman*, which uses the growth of the individual as synecdoche for the progress of civilization, Wells's last human novel can be said to exemplify the *unbildungsroman*, advancing through the reduction, diminishment, or de-structuring of character to the point of its existential collapse.[32]

Narrating extinction, thus, becomes sensible through its existential effects on characters and on the literary device of character. Wells then absorbs the existential ends of character and the limit ends of other devices back into the novel, rebuilding the novel essentially from its own ruins. In Wells's hands, both biology and the devices of narrative and character become atavistic and futuristic at the same time. Such speculative atavism at the level of the form and content of character remarkably provides a launching pad for the new genre of science fiction. This genre will already have incorporated extinction and the finitude of narrative as its uncanny enabling fictions and its mythological origin and ends. At the same time, this foundation precipitates its own unraveling—the destroyability of character, which suddenly surges up in Wells's narrative, becomes the evolutionary conditions of character from which individuals cannot escape.

It is a common lament that utopian fictions often treat individual characters perfunctorily, as this genre emphasizes the collective scale details of world-building and ideal social organization. Utopian characters can be flat for the narratological reason that the social system takes precedence over individual lives. Dystopian novels, which feature the dismantling of worlds and social organization, often whittle character down to levels of survivalism, an issue I will return to shortly. But last human novels feature flatness of character in part because the status and function of character itself is reaching a terminus. The Time Traveller's response to the denouement of the species defines him more than his exploration of his own subjectivity. However, it is via the presentation of the existential limits and the flatness of character and the narratological devices of time travel into deep time that allow the novella to catch a glimpse of the totality of the human species.

Reading the formalist aspects of Wells's novel as conjoined to its development of a literary culture of extinction re-establishes the importance of previous genre-theory approaches to the novel. Patrick Parrinder's long-standing observation that *The Time Machine* "responds to the breakup of the coalition of interests in mid-Victorian fiction" by eschewing both realism and utopianism remains revelatory as it points to the social demand for a new genre as well as indicating the effect that the incorporation of extinction into the novel had on the coordinates of narrative. For Parrinder, Wells's novel "avoids certain limitations of both the Victorian realist novel and the political utopia. An offshoot of Wells's use of fantasy to explore man's temporal horizons is that he portrays human nature as at once more exalted and more degraded that the conventional realist estimate."[33] Wells's vast temporal arc and his heavy use of extrapolation to imagine the traumatic undoing of human character served as critical counterpoint to the faltering bourgeois-realist emphasis on humanistic "roundness" of character and plot (and perhaps this is the moment to point out that all characters are "flat" in the sense that, bound to the page, they will never become alive).

By imbricating the genre of scientific fantasy with the extinction plot, Wells assumed a distance from middle-class self-interested moral constraints and made room for promoting a futurist "nostalgia for the present"[34] that he understood as requiring masculinist traits of sturdiness, literary cleverness, and "carrying on" in the face of species-wide precarity. Darko Suvin, who notably declared Wells's *The Time Machine* as "the turning point in science fiction," remarks that Wells's "satisfaction at the destruction of the false bourgeois idyll is matched by his horror at the alien forces destroying it."[35] Biological and planetary entropy are the "alien forces" in *The Time Machine*, but Wells's subsequent novel *The War of the Worlds* will call upon "real" aliens from Mars, an exoplanetary extinction threat that will be defeated precisely as British bourgeois norms of fortitude and pluck can be reclaimed for species survival.

For Suvin, Wells's first novel offered a paradigmatic case of the argument that science fiction is a "literature of cognitive estrangement." The Time Traveller is not only a figure for estranged cognition; he also confronts a vision of the end of cognition. Joanna Russ commented that Wells's novel helped contribute to the genre of creative downfall, what she calls "the unfortunate rise" whereby "the very production of intelligence, of mind, is what must, sooner or later, destroy mind."[36] And yet readers can consume the destruction of cognitivity ironically as a cognitive enhancement, the theme of so many of Wells's novels.

*The Time Machine*, ultimately, does not have to resolve the formal paradox and existential trauma of presenting a point of view for a world that has lost the possibility of any point of view, because the time machine allows the Time Traveller to return to his Victorian company and tell his tale in a space and time where storyworlds are not in crisis. We are brought to the edge of the annihilation of Earth and to a point where plot has nowhere to go—yet, then, we are immediately brought back into the bourgeois, well-heeled, class-stratified, colonialist, steampunk, all-male interior of the dinner party.

"No. I cannot expect you to believe it," states the Time Traveller upon his return to his fellow witnesses in another meta-fictional moment that turns on the self-negational properties of fiction. "Take it as a lie—or a prophecy. Say I dreamed it in the workshop. Consider I have been speculating upon the destinies of our race until I have hatched this fiction. Treat my assertion of its truth as a mere stroke of art to enhance its interest. And taking it as a story, what do you think of it?"[37]

Of course, this *is* a fiction; the Time Traveller is trying to call their bluff while disclosing the time travel paradox that human extinction will always be a fiction and an imaginary future until the event occurs that renders the difference between imagination and reality mute. In the paradoxical phrasing of Jean-Luc Nancy, "Literature does not come to an end at the very place where it comes to an end."[38] The Time Traveller's admission of fiction suddenly deflects from the novel's utterly anti-human vision of extinction, annihilator of all bourgeois values, including the value of fiction. Yet even the extinction of the human can be humanized and can bolster humanistic norms, as the initial narrator returns to offer a moral epilogue: "He [the Time Traveller], I know . . . thought but cheerlessly of the Advancement of Mankind, and saw in the growing pile of civilization only a foolish heaping that must inevitably fall back upon and destroy its makers in the end. If that is so, it remains for us to live as though it were not so. But to me the future is still black and blank" (91). The future is still metaphorically a blank page that will require writing with ink into the black.

## Literature after Survivalism

Here, we return back to the chapter's initial question: Why do readers want to read about their own extinction? Why do readers want to encounter over and over—given the widespread popularity and formulaic aspects of the last human genre—visions of the end of the world? Extinction narratives simulate a feeling of death but, at the same time, also a feeling of life—these are existential extremes of what it means to feel like a species. This feeling is a version of Rousseau's statement in *Confessions*: "I can very well say that I did not begin to live until I looked at myself as a dead man."[39] Wells's view of the end of the world, ultimately, is quite compatible with the point of view of the everyday bourgeois life of the novella's readers. These readers are invited to imaginatively engage both the end of the world and the stream of everyday life together. Only both perspectives combined—the present and deep time—will reveal the aesthetic and scientific possibilities and perils of extinction. In the meantime, species elegy combined with the literary possibilities of scientific romance can provide coordinates for furnishing future visionary experiences in the present.[40]

Wells, in his preface to the 1931 edition of *The Time Machine*, rather cheekily disparages the quality of his youthful writing but notes "certain originalities in it saved it from extinction."[41] Extinction and invention, extrapolation and negation, the end of humans and the birth of new scientific endeavors to expand human consciousness are juxtaposed and co-enabling conditions for Wells's fiction. Seeing the world from the point of view of the end of the world is a means to begin to comprehend phenomena of deep time, long-scale geological and biological processes, and cosmological entropy as realities both existential and formalistic for narrative. All of these are prime material for transforming the scope of the novel at the turn of the century, because they press readers to extrapolate the evolutionary trajectory of the human species at the same they explore the ruinous futures of narrative, character, and point of view.[42] To see character from the end of character, and plot from the end of plot, provides the reader a special way to comprehend both the work of fiction and a world without fiction.

I want to close this chapter with some final thoughts about the effects extinction has on literary form and the readers of this genre in the present moment. I have noticed recently in teaching dystopian fiction that many readers gravitate to what can be described as reading for human survival. I see many student readers combining a textual experience with what I would call a prepper experience or mentality. They find dystopias read less like allegorical fantasies and more like planning manuals. Readers are identifying with the existential crises of plot and form and mimetically assuming they are reading for their own existential survival. They are reading dystopian fiction as though such works provide a nonfictional and thoroughly plausible script for how to survive an apocalyptic event—as if the most genuine way to read these works is for non-aesthetic practical preparedness.

There are many different survivalist or prepperist plots in recent apocalyptic fiction—including writing by Octavia Butler, Cormac McCarthy, Colson Whitehead, Margaret Atwood, and Emily St. John Mandel, among many others—yet, by reading these very different texts for the sake of survivalism, the diversities of style, formal range, social nuances, embodied perspectives, and the experiences of alternative worlds of fiction are often streamlined into a thin motivation to outlast the end of the novel. Students find themselves increasingly familiar with literary extinctions and increasingly sure of the need to read for survival.

This mode of reading now extends well beyond dystopian fictions. Students are finding that reading for survival can be extended to any literary work, as well as to broader contemporary conditions such as the state of the humanities, humanistic ideals, and much of current political phenomena. A significant number of recent video games, including the wildly popular first-person shooter game *Fortnite*, have as their aim for the player to be the last one standing. To merely survive is to win. Reading (or playing) for survival is becoming reading (or play) as such. First-person survivalist narrators become fused with first-person/last-person shooters. First person becomes last person—no matter what one is reading or playing or doing. Thus, "last personhood" is becoming more common as the basis for personhood generally.

When we talk of literatures of extinction today, we need to talk about the rhetoric of survivalism and prepper mentalities that are increasingly becoming attached as primary modes of everyday thinking as well as responses to situations of extinction and planetary risk. This is the case even when the past two centuries of the literary culture of extinction, with its frequent foregrounding of last humans, is evidently so different from the current ecological reality of extinction, where humans are the peak species at peak numbers while a sweeping range of animals, along with many marginalized human communities, face an existential precipice. Yet the fictionalization of extinction into a genre, even as it appeals to the reader-as-survivor, trades contrariwise on literature's suspension of the reality principle and the daily pressures of self-preservation. Survivalist readers who mimetically identify the form and content of extinction with their own reality are especially missing how the aesthetic contradictions of last fictions also permit readers to distance themselves from their immediate world.

As Theodor Adorno repeatedly stressed, there is a dialectical phenomenon embedded in literary form, such that the literary plot of survival brushes against the experience of reading in which one is able to suspend the immediate world and its ruthless demands for nonstop self-preservation. An artwork's form requires some creative distance and visionary alterity from already existing reality. In Adorno's view, "The concept of form marks out art's sharp antithesis to an empirical world in which art's right to exist is uncertain. Art has precisely the same chance of survival as does form, no better."[43] What Adorno means by art's survival is enigmatic, but we can say that art's persistence (along with the survival of fictional characters) is not reducible to any direct representation in the work of an instinct for self-preservation. Because "form is mediation" (144), even literary dystopias are utopian in form because to read them requires we remain as readers, not destroyers of language. Furthermore, if one has time for fiction, one is not immediately focused on maintaining the brute facts of one's life. To read is to maintain a margin of freedom and disinterest from necessity, utility, and ravenous appetites. Adorno also recognizes that the dissolution or negation of art has a peculiar aesthetic attraction that, paradoxically, is tied to the ongoing promise of art as such: "Daringly exposed works that seem to be rushing toward their perdition have in general a better chance of survival than those that [are] subservient to the idol of security" (177).

Overly formulaic art—which can include now-tired apocalyptic genres—expires quickly and ironically because it reductively strives too hard for survival. However, lest one see artful perishability as the straight gate to aesthetic longevity, Adorno adds: "Speculating on survival by adding something perishable is hardly helpful" (177). Adorno also stresses that the survival of the work of art as an object retaining aesthetic resonance over time—which is not the same as the expression of survivalism at the level of its content—is predicated on its ability to embed in its formal constitution an experience of socially shared freedoms, a release from the demands of self-interest and coercion, a reflexively aware sensuality, and an unalienated relationship to nature not predicated on the need to fetishize either nature or civilization. All of which, if realized, would portend a different kind of "end" of art than the version in which art finds its end in last human scenarios.

Reading the literature of extinction should raise questions of how the survival plot shapes literary form and content, and such narratological investigations do not reduce themselves to the singular question of who or what lives on. There is a critique of the survival plot inside the survival plot—having the time and imaginative capacity means the reader is not facing the dire demands of immediate self-preservation and, thus, experiences an anti-apocalyptic aesthetic at the formal level. The aesthetic pacing of the text, even if it tells the story of survivalism, offers suspension from the demand for constant survivalist mentalities and opens readers toward the project of imaginative worlding.

How can reading survivalism at the level of form and content contribute to a broader ecological knowledge, care, and relationality, what Wai Chee Dimock calls "assisted survival"[44] in and through narrative connectivity? Noel Gough remarks that "one of the most significant and distinctive contributions that SF has made to environmental literature has been to represent the endless variety of ways in which life on earth can be destroyed."[45] But Gough does not see this world destruction as necessarily role-playing for dystopianism, and points to how such apocalyptic literature can demonstrate environmental renewal and imaginative acts of alternative creative worlding. The Time Traveller's own meta-literary commentary on the status of his narrative enjoins his parlor room listeners to evaluate as a collective the effect of encountering the end of all storyworlds. Hence, one need not read the human extinction plot or survivalist plot strictly from a survivalist or prepperist subject point of view—or at least, the prepperist reading is just one reading, not the ultimate reading or last reading. Survivalism is just one kind of last trope, and not necessarily *the* last trope.

In the case of Wells, there are a number of ways *The Time Machine*'s linkage of survivalism to masculinity and biopolitics allows for certain kinds of reactionary prepperist readings of extinction to take center stage. The eugenic "we" of this novel is white urban British men who see themselves variously reflected in the Time Traveller, narrator, and post-human Eloi and Morlocks. To come back to the question of why Victorian readers would want to pay to read of their own extinction, in addition to the pleasures of encountering the existentiality of literary form, such a reader also has the privilege of seeing their own *fin du siècle* as *fin du monde*. That is, readers have the pleasure to survive themselves and nominate themselves as figuring the end of the human, if not humanity's saviors. They are victim but also victor, because there is no one to defeat them.

Perhaps this is why many theorists have noted recently that it is difficult to convince readers of dystopia to also be committed readers and practitioners of utopian ecological conservation. Readers of dystopia and extinction are enabled by the fantasy of the survivalist genre enduring the end of genres. Like the Time Traveller, these readers consume their own formal and existential limits. Reading for survival has no need to read for *bildung* when *unbildung* radiates everywhere. Survivalists fantasize themselves as last readers who do not have to make room for other readers.

But reading for survival is not the only way of reading through one's own extinction narrative—many initial readers saw Wells's text as satire and an allegory of changes in genre that ushered in the rise of scientific romance rather than having any socially mimetic, visionary, or predictive power. Wells himself came to read his story as allegory rather than realism, reflecting on it later:

> It was still possible in *The Time Machine* to imagine humanity on the verge of extinction and differentiated into two decadent species, the Eloi and the Morlocks, without the slightest reflection upon everyday life. Quite a lot of people thought that idea was very clever in its sphere, very clever indeed, and no one minded in the least. It seemed to have no sort of relation whatever to normal existence.[46]

Wells views his work less as a lesson on the imminence of extinction and more as an exercise in thinking creatively and inventing new genres that reflect on what it means to be a species. While formulaic narratives of ecological decline and apocalypse remain a concern because they bespeak a pattern of hopelessness, perhaps just as alarming are these increasingly entrenched ways of reading for survival in today's dystopian and last human narratives.

# 4 /   Concepts of Extinction in the Holocaust

This chapter inquires into the concepts of extinction circulated and acted upon in Germany at the time of the Holocaust. Nazi ideologues appropriated the scientific knowledge about extinction developed in early twentieth-century biology and the emerging field of ecology to propagate the notion of a "state biology [*Staatsbiologie*]," a term for the fusion of politics and biology used by the biologist Jakob von Uexküll in 1920 and picked up by Nazi officials. The political biology of the Third Reich promoted a relentless watchfulness over extinction risks, not for the human species as a whole but for the German people only. Confronting extinction became not a minor concern or remote problem but, rather, something of the utmost importance in daily life, centrally informing all decisions small and large. Stated most directly, the Nazis were obsessed with ideas of extinction, viewing the decline and decimation of races and species as having an ultimate causal power and casting definitive judgment across all of life.

The Nazis conceived of extinction generally in two ways: as a biological-based identity crisis facing the German people in the near future if they did not adhere to racial purity, and as a biopolitical force that could be applied to humans and animals to protect Germans and to Germanicize the Central and Eastern European landscape. They came to see extinction as both a rallying idea to unify the Reich and a practical tool to dispense with any perceived opponent. The Nazis feared for their own imminent extinction and presumed that all other peoples across the globe could be rendered extinctable if necessary.

The extinctionary agenda of the Reich certainly is already well known and heavily researched; what is less understood is the contextual history of the range of concepts of extinction and animal finitude that informed this agenda. Upon taking power in 1933, at a time when the German population set a new peak at 65 million inhabitants, the Nazis claimed that German "Volk" life was facing imminent extinction if the German nation did not act to correct its own biological downfall, an argument supported by some of the most knowledgeable biologists in Germany at the time. The Nazis sought the eradication of Jews, the Sinti and Roma peoples, as well as physically and mentally disabled persons, all of whom were blamed for precipitating a biological degeneracy that would lead to the downfall of the Volk. Eliminating such peoples supposedly would be the means to achieve a pure Aryan race that was fantasized as non-extinctable. The Nazis created an entire way of life based, in significant part, on a set of pseudoscientific and delusional ideas of extinction as applied to themselves and to everyone who was not themselves. The Third Reich sought to change radically the definition of extinction across the globe and to destroy the means by which to witness and judge such ultra-violence. The legacy of this nightmare still has effects on the meaning of extinction today.

Nazi ideologues, rank-and-file members, and the many collaborators, associates, and beneficiaries of the destruction of European Jews did not always consistently define and determine the scope of their attitudes toward an agenda of genocidal extinction as the means to rid Germany and its surrounding territory of peoples who were declared biologically unfit. Historians have sought to demarcate when, exactly, the Nazi regime decided on the total extinction of Jewish lives across the globe rather than policies of local segregation and disenfranchisement with the goal of displacement.[1] The timing and degree to which devotees to Nazism believed in and acted on an eliminationist agenda against the Jews already has been extensively researched.[2] My purpose here is to examine how the Nazis drew on and manipulated concepts of extinction drawn from biology, racial eugenics, medicine, ecology, and perceptions of animals. The Reich's attempt to redefine extinction—based on a set of scientific and medical notions cherry-picked from a handful of natural science texts mixed with academic falsifications, mythological fantasies, and pathological obsessions regarding purity—formed a significant contribution to their claims for racial and biological superiority. In addition to the fascist system predicated on "organized lying" (Hannah Arendt)[3] and "organized cruelty" (Bruno Bettleheim),[4] the Nazis also pursued an agenda of *organized extinction* in their attempt to remake biological life in conformity with a political doctrine of Aryan supremacy. I also discuss in this chapter the nascent effort during the Third Reich to revive extinct animals perceived to have a Germanic lineage as enmeshed in these efforts to redefine and control species finitude. Finally, I examine on how biopolitical theories developed by Hannah Arendt, Giorgio Agamben, and Roberto Esposito of the Nazi attempt to obtain power over life and death must include a further reflection on how ecological references to species precarity, in contrast to the attempted redefinition of extinction, provided a glimpse of an alternative world outside Nazism for the persecuted.

Genocides do not need to be based on an elaborate definition (or attempted redefinition) of extinction. Although the cases of genocide in Rwanda perpetrated by Hutus against the Tutsis and in Turkey by the Turks against the Armenians involved assertions of the superiority of one culture against the other, the primary aim was to obtain land, enforce ethnic or cultural separation, and remove the existing inhabitants. The Nazis, however, had an elaborate argument for the centrality of a new science of extinction in their *Weltanschauung*, and they placed the highest authority in biologists, doctors, anthropologists, zoologists, and political philosophers who had claimed to study the natural and cultural history of extinction. The Nazis sought to impose a new global doctrine on the meaning of extinction—that the entire world should be reducible to matters of extermination versus purity—and that a new biological hierarchy (expediently confirming their own racial supremacy) should be the outcome of such a doctrine.

To be clear, the Nazi devotion to placing extinction at the center of their political program does not equate to a monocausal theory of Nazism or the Holocaust. The destruction of social and legal norms in Germany and its occupied countries, the erection of ghettos and concentration camps, and the establishment of totalitarian rule in the name of a pseudo-scientific vision of Germanic superiority certainly require attention to complex motives and multiple explanations. Coursing through these major political agendas was a belief that the matter of extinction could be at stake to some degree in all social relationships and civilizational projects. In Nazi totalitarianism, the extreme poles of either race purity or extinction were proclaimed to be of paramount importance in every action or decision, no matter how apparently unrelated the instance might be to anything evidently concerning the biology of species and populations. The constant drumbeat of public rhetoric declaring an extreme biological crisis facing Germany—despite any evidence of reality—required constant repetition across all aspects of daily life.

Adolf Hitler made it clear in *Mein Kampf* that he thought the entirety of the German people might soon become extinct, using the established terms *Aussterben* and *Ausrottung* in biological discourse for the end of a species and applying these to national and racial groups.[5] In his view, to lose one's purported racial purity was tantamount to extinction. Hitler claimed that every Aryan had to act keeping foremost in mind the "condition in which each individual knows that he lives and dies for the preservation of his species."[6] According to Hitler's argument, the unique biology and culture of the German Volk could be saved from perishing only through a policy of racial purity by "exterminating" other humans deemed "germs." It is well known that Nazi propaganda cast Jews as lowly, contagious pests and vermin (epithets used across Hitler's writings), which contributed to the development of the idea that they were deemed subhuman and extinctable in advance.

The treatment of Jews as extinctable life opens onto the issue of what the conjunction of notions of extinction, the actual lives of animals, and the concept of a Germanic nature meant at the time of the Third Reich. National Socialists proclaimed that racially pure Germans, as well as regional animals and the landscape itself, formed a biotic unity, an *"organische Lebensgemeinschaft"* (organic life-community). The dictum of "blood and soil" asserted Germanic racial family lineage and historically Germanic lands as the essential standard of value in the world and the basis for assessing and hierarchizing all life and land across the planet. The Nazis hierarchized animals according to similar principles, where purity of breed, nativity to Germanic-associated lands, perceived historical importance and nobility, and robustness of stock would rank as best. Animals that were of mixed descent, from non-Germanic lands, deemed ugly, pestiferous, or weakened by domestication were of lesser importance and, ultimately, destroyable.

The Nazis, thus, did not divide the world into humans and animals (condemning Jews to the status of animality) as much as they divided the world into Germanic humans and animals and non-Germanic humans and animals. This schema divided biological existence along the lines of Germanic life deemed as absolutely valuable and non-extinctable compared to non-Germanic life, which was subordinate and could be extinctable in advance. In the phrasing of one Nazi party member, "The animal kingdom is the model for the organic state of National Socialism, the 'biotic community,' or still better, the 'living space' [*Lebensraum*] and our 'vital society.'"[7] Dedication to the Germanic biotic community above all else rendered other human and nonhuman lives subordinate or dispensable. In Ehrhard Wetzel's "Comments on the General Plan for the

East," the genocidal proposal to clear the peoples of Eastern Europe for the expansion of German *Lebensraum*, he cited professor of anthropology Wolfgang Abel's view that "saw only two possible solutions: either the extermination of the Russian people or a Germanization of its Nordic elements."[8]

The utter disposability of non-German life appears most evidently in Heinrich Himmler's notorious speech at Poznań on October 4, 1943, where he declared:

> We must be honest, decent, loyal and comradely to members of our own blood, but to nobody else. What happens to a Russian, to a Czech, does not interest me in the slightest. . . . Whether nations live in prosperity or starve to death interests me only so far as we need them as slaves for our culture; otherwise it is of no interest to me. Whether 10,000 Russian females fall down from exhaustion while digging an anti-tank ditch interests me only insofar as the anti-tank ditch for Germany is finished."[9]

Himmler added that he did not condone violence when not "necessary," and insisted "We Germans, who are the people in the world who have a decent attitude towards animals, will also assume a decent attitude towards these human animals. But it is a crime against our own blood to worry about them and give them ideals."[10]

This chapter can only begin to discuss some of the complex and volatile reasons behind associating Jews and other undesirables targeted in the Holocaust as biologically pernicious and, therefore, eradicable. In recent years, scholars have brought animal studies and environmental history methods to the study of the Holocaust to understand better the implications of declarations such as those made by Hans Schemm, who stated in 1929: "National Socialism is politically applied biology."[11] Examples of this research include work by Boria Sax, Charles Patterson, Frank Uekoetter, Thomas Lekan, Roberto Esposito, and Timothy Snyder, among others.[12] Recent scholars also have emphasized the need to develop an ecological historical approach to studying the Holocaust.[13] These works offer important reasons to comprehend the multiple, entangled, distorted, and inconsistent associations of humans and animals in Nazism. The Nazis passed more than a dozen laws mandating the prevention of cruel treatment of animals in domestic care, farming, and scientific research while also instituting laws that limited hunting, extended protections to wildlife, and codified wild habitat preservation. An argument can be made that National Socialism saw the most wide-ranging local ecological and animal protection laws passed compared to any other nation at the time;

however, these protections were consistent with the idealization of Germanic nature and the domination of other lands and peoples and, thus, perpetuated a broader form of environmental violence (now identified as ecofascism). The Nazis also supported research into the possibility of reviving previously extinct animals identified as Germanic. Extinction and de-extinction could both serve to Germanify life. The most prominent work in de-extinction along these lines was led by the zoologist Lutz Heck, who helmed an effort to bring back the aurochs, which he justified in biomythological terms suitable to the Reich.

Nazi ideologues employed several terms for extinction taken from academic biology and from wider global public awareness of the mortality of species. Biological scientists and textbooks in German prior to the Third Reich used the term *Aussterben* as equivalent of the English "extinction" and *Vernichtung* for "extermination." Darwin used both *extinction* and *extermination* as generally interchangeable terms in *On the Origin of Species*. In this book, Darwin did not connect either term to intentional human eradication of other species or other humans. However, as mentioned in chapters two and three, Darwin specifically used the term "extinction" in *The Descent of Man* to apply to historical examples—including Western colonial conquests—of one group of people taking the territory of another and effectuating the displacement, absorption, or destruction of the previous peoples. The avid Darwinian and highly respected German biologist Ernst Haeckel, in his *Natürliche Schöpfungsgeschichte* (published in1868; translated as *The History of Creation*, 1876),[14] used both *Aussterben* and *Vernichtung* (in both noun and verb form) to indicate extinction or eradication of species. Haeckel also employed another term, *Ausrottung*, which can be translated as extermination or eradication, on three occasions. The first use is to mention that weeds are typically eliminated in gardens (155), and then Haeckel immediately reuses the term in defense of the death penalty for unrepentant criminals whose biological erasure, Haeckel claims, would be a benefit for society. A third usage of *Ausrottung* applies the term to the eradication of Indigenous peoples in the Americas and Australia and sets their mass deaths in the context of Darwinian natural and artificial selection (739).

The Nazis used these terms (Hitler used all three in *Mein Kampf*) as well as others in developing their plans to ultimately eradicate Jews while also using the same terms as applicable to Nordic or Aryan people whose loss of supposed racial purity would be seen as equivalent to extinction. These are two radically different usages of the concept of extinction—one to promote the murder of a people and the other to claim that loss of racial purity would be the equivalent of the disappearance of a race or subspecies—yet the Nazis intentionally did not see a difference. The Nazis also did not distinguish among terms for extinction or eradication taken from a biological lexicon and terms from a medical lexicon for the eradication of diseases and general antiseptic practices. One Nazi party member, in a speech on the opening of the Dachau concentration camp in 1933, insisted that the inmates, particularly political prisoners of Jewish descent, must not just "die" (*sterben*) but should go "extinct" (*aussterben*): "*Sie brauchen nicht zu sterben, aber aussterben sollen sie!*"[15] An anonymous anti-Nazi publication that appeared in 1936 under the title *Der Gelbe Fleck: Die Ausrottung von 500,000 Deutschen Juden*—which collected the sayings and evidence of the National Socialist Party's murderous agenda against the Jews—highlights in its title the use of "*Ausrottung*" as a term for extermination.

The Nazis wanted academic and scientific justification for their redefinition of extinction. As several historians have noted, one of the unique aspects of the Holocaust was the degree to which many university professors and students became early supporters of extreme racial and biological "redemptionist anti-Semitism" (Saul Friedländer),[16] the notion that the elimination of the Jews would somehow have a messianic impact on the life of Germans. Hitler stated in one of his monologues during the war: "We will become healthy when we eliminate the Jews."[17] Numerous academic biologists in Germany during the Third Reich, who certainly bolstered the notion that Germanic peoples constituted a distinctly identifiable race and a unique biology, also gave scientific credence to the argument that the German *Lebensgemeinschaft* could become extinct in the same way that a species wholly disappeared. Ernst Lehmann, a professor of botany and appointed in 1931 as chairman of the German Association of Biologists, clamored for the salvation of German-ness at the same time he applauded the "cleansing" of "foreign races." Lehmann situated "Germanic" biology itself as redemptionist, claiming: "Today, in view of the extreme existential affliction of our Volk, there arises before the German biologists as his most pressing and greatest task that of sharply formulating the laws of life from which our German Volk lives and from which the National Socialist *Weltanschauung* is born."[18]

Such arguments treated terms of life, species, and race as essentially fixed and rooted in place by ethnicity and nationality. By contrast, Darwin stated he could not strictly define a species as compared to a variation, and insisted that a precise definition was not needed. Darwin's emphasis on entanglement, adaptation, change, non-teleological development, and chance factors prompted his keeping of definitions of species open-ended. Moreover, Darwin's reflections on extinction did not lead him to ascribe a winner-take-all scenario in which just a few species dominated all the rest. Darwin understood his model needed to explain how there could be both processes of extinction and an "entangled bank" teeming with biodiversity and a plurality of species differences.

National Socialist biologists treated the terms of *extinction* and *selection* extracted from Darwin's ecological contexts as immediate political and medical directives. They focused especially on the notion that distortions in nature created by artificial selection—associated with the invasiveness of Jews, but also with city life and domesticity—had disrupted the supposedly healthier and more holistic processes of natural selection. The aim was paradoxical in that the German state had to use artificial selection by removing "foreign races" to restore an archaic and mythical realm of Germanic nature championed by party ideologues.

The selective elimination (*Ausmerzung*) of some races and the obsession to physically and spiritually improve the Aryan race became the central doctrines of *rassenhygiene*. The Nazis subsumed an array of modern biological concepts that were methodologically and conceptually open to multiple causes and situations into a movement focused on marking all existences and achievements according to demarcations of race purity destined either for immediate salvation or extinction. For example, the noted biologist Jakob von Uexküll, in his book *Staatsbiologie: Anatomie, Physiologie, Pathologie des Staates* (first published in 1920), added a new chapter in 1933 on "parasitic diseases," designating those of "alien races" as parasites. Uexküll had not joined the NSDAP but confidently wrote, "No one will criticize the leader of the state when he puts a stop to the infiltration of the organs of the state by an alien race."[19]

Robert Proctor analyzes what he calls the "medicalization of anti-Semitism,"[20] the notion that the Jewish people as such were a medical problem (supposedly inducing illness in non-Jews by their mere presence) requiring a medical solution, for which Hitler presented himself as the supreme "doctor." The Nazi valorization of race purity and vigilance toward extinction thus involved an imbrication of biological, medical, and political terms across multiple disciplines, blending scientific etymologies with laypersons' uses to the extent that one could no longer discern exactly which concept came from which discipline. Each of these terminological sources contributed to definitions of extinction directed outward toward non-Germans and inward toward efforts to assure a Germanic purity that was perceived to be on the brink of survival.

Konrad Lorenz, a young Austrian biologist when he joined the Nazi party, who later won a Nobel Prize in physiology (1973), made several statements connecting the supposed precarious condition of the German race with imminent extinction. Lorenz was not alone in this,[21] but considering his knowledge of Darwinian biology and his appointment as a professor of animal psychology in 1940, Lorenz had justification to claim substantial authority for his ideas. In an article published in *Der Biologe* in 1940 (a respected academic journal established in 1931; by the late 1930s, several NSDAP party members had taken central roles in editing and publication of the journal), Lorenz argued that race, rather than the human species in general, was the primary biological unit under selective pressure. Lorenz conjured a vision of the German race as threatened in the very near term with extinction, and analogized this potential extinction event with the most obvious reference of extinction, the dinosaurs:

> Whether we share the fate of the dinosaurs, or whether we raise ourselves up to a higher level of development, undreamed of and perhaps inconceivable with the present organization of our brains, is exclusively a question of the biological penetrating power and life will of our *Volk*. In particular, the great decision most likely depends on the question whether or not we learn in time to combat certain symptoms of decline that originate in lack of natural selection. In this race for existence or non-existence, we Germans are a thousand strides ahead of all other civilized peoples.[22]

Lorenz claimed that Germans as a race were in a better position to survive than most others, but warned of "degeneration, through the racial and moral decay caused by big-city life, declining birth rate, carcinoma and world capitalism and countless other forces hostile to the *Volk*."[23] Lorenz, with his advanced knowledge of modern biology, assessed the German Volk with the conclusion that their extinction, like "the fate of the dinosaurs," might be a near-term possibility.

Germany was by no means on the precipice of civilizational collapse or biological downfall at the time, even taking into account setbacks to prosperity from losses in World War I and the Depression of the 1930s. The Reich's emphasis on extinction, however, did not distinguish between metaphorical and non-metaphorical, or rhetorical and scientific, uses of such terms. In arguments such as Lorenz's, Germans are to think of themselves as potentially soon meeting the fate of the most iconic animal for extinction, the dinosaurs—whose survival as a collective of species for several hundred million years seemed inconsequential since the species has become synonymous with disappearance. Lorenz exhorted Germans to view the world as bifurcated in the metaphor of running a race that will end either in existence or non-existence.

From biological scientists to political party leaders, the Nazis encouraged an inflationary, hyperbolic, and ubiquitous use of the terms of extinction. Such use did not allow for nuanced methodological distinctions and debates and, instead, facilitated the collapsing of different scientific and cultural frameworks regarding species finitude. The ubiquitous rhetoric of the supposed imminence of German extinction made it possible to find the perils of extinction in any aspect of everyday life. The omnipresent concern for extinction supposedly required enforcing a regime of racial "health" and biological "security" that did not distinguish between actual bodily illness, explicit propaganda, and manufactured hysteria. The Nazis conflated biology and myth, natural history and conspiracy, in their insistence that the immediate situation in Germany could result in only one of two outcomes: biological downfall or biological utopia (Lorenz's "higher level of development, undreamed of"). They made no distinction between the deep-time natural history of extinction and ideologically motivated murder, or between ecological history and hell.

The Nazis believed they could regenerate themselves through the extinction of others. They wanted eternal death for non-Germans to translate into eternal life for themselves. The Nazis also sought to destroy the means by which to measure the difference between the ecology of extinction within natural selection and murder. If they could, the Nazis would have destroyed the combined empirical and ethical bases for the science of extinction and replaced it with organized pseudo-scientific fabrication attached to political axioms of dominance. Whereas Darwin noted in *On the Origin of Species* his "metaphor" of the "struggle for existence" "included dependence of one being on another"[24] and commented on the "beautiful co-adaptations" that included "the humblest parasite which clings to the hairs of a quadruped" (51), the Nazis viewed racial struggle and extinction as a winner-take-all phenomenon that promised the ruination of any sense of commonality and co-adaptivity. Though the Nazis claimed to be acting according to the supposed dictums of the Hobbesian and Darwinian state of nature, as Terrence Des Pres comments, "The 'state of nature,' it turns out, is not natural. A war of all against all must be imposed by force."[25] It was delusional human malice, not acting in accordance with nature or the biological laws of extinction, that created the concentration camps; as Tzvetan Todorov remarks, "torture and extermination have not even the remotest equivalent in the animal kingdom."[26]

The Nazis claimed they wanted to live by biology and nature—supposedly static laws rooted in the soil that could not be debated or changed—but, really, they wished to live by their definitions of Germanic biology and nature while commandeering scientific authority. They understood that extinction, as an extreme and traumatic phenomenon that loomed over any species, could be mobilized in public discourse with the effect of collapsing and undoing the epistemological and psychic frames used to understand it; in effect, traumatizing the German people so they could not tell the difference between extinctionary rhetoric and evidence-based evaluation of biological perils. Exploiting such traumas involved destroying the basis for recognizing what constitutes a traumatic event in the first place, redefining mass murder and extinction of others as having no traumatic standing.

Perhaps most disturbing of all is that the Nazis were able to invent something worse than extinction, for which the new term *genocide* was coined. Genocide included eradication with humiliation, forcing the victims to participate in their own annihilation, blaming the victim, and seeking the victim's erasure physically and metaphysically, destroying the physical body as well as the very idea of a group of people. The Nazis sought to use politically-motivated extinction to go beyond and render incoherent the natural history of extinction, practicing what Saul Friedländer calls an "amorality beyond all categories of evil."[27]

## Endangered Animals and Animalities in the Third Reich

The omnipresent and conspiratorial use of concepts of extinction in the Third Reich raises the question of how this term was applied to the actuality of endangered animals in the Third Reich. The Nazi fixation on concepts of extinction occurred at a time when there was a noticeably significant reduction in biodiversity in the country, with a marked dwindling in numbers of wild animals, while the numbers of domesticated animals increased at a rapid pace. Many of the animals identified as important to Germanic nature, such as grey wolves, brown bears, and eagles (golden, spotted, and white-tailed eagles live in small numbers in Germany; the national coat of arms features an eagle that does not specifically correspond to any of the native eagle species), had been depleted significantly or extirpated over the previous centuries.

As the Nazis fanatically obsessed over what they perceived as the imminent demise of their proper "organic life-community," how did the status of actual animal populations contribute to the concepts of extinction that circulated during the Third Reich? What did extinction and the Aryan attraction to Germanic animal life mean, then, in a time and place that had already undergone several extirpation events and seen dramatic changes in the lives of animals?

Germany in the 1930s was not alone among European countries to confront a dramatic drop in wildlife populations. Much of Western and Central Europe already had been noticeably depleted of wild animals by the end of the nineteenth century in what David Blackbourn calls "a war against untamed nature," leading to the "culling or eradication of creatures that directly challenged humans for resources."[28] In Germany, the change in hunting laws in 1848 opened the hunt more widely to all landowners and, consequently, led to a rapid diminishment in wild animal numbers. The grey wolf was extirpated from Germany at this time.[29] No wolves dwelled in Germany during the Third Reich, yet Hitler "enveloped himself, both personally and militarily, with wolfish terms."[30] In *Animals in the Third Reich*, Boria Sax suggests that the lack of actual wolves served to facilitate the animal's mythologization (Hitler's code name was "the wolf"). Even though wolves already had been decimated, Sax adds: "The Nazis were constantly invoking dogs and wolves as models for the qualities they wanted to cultivate: loyalty, hierarchy, fierceness, courage, obedience, and sometimes even cruelty."[31]

Brown bears, which historically existed throughout Germany, disappeared in the country in the nineteenth century due to deforestation and hunting.[32] The Eurasian lynx and Eurasian beaver also were hunted to near extirpation in Germany into the early twentieth century. Avian populations plummeted from the eighteenth to the twentieth centuries, with some hunted as game while others were treated as pests that attacked crops, such as sparrows, which were killed in the millions during that period. An extinctionist attitude toward sparrows, raptors, and shrikes prevailed as late as the end of the nineteenth century. Hans von Berlepsch, in his uncannily titled *The Complete Bird Protection* (1899), extolled the "extermination of the different enemies of the birds worthy of protection."[33]

At the same time many wild animal populations collapsed, Germany's agricultural production and animal domestication industry was booming at the outset of the twentieth century. Cattle increased in number from approximately 10 million in 1800 to 21 million, according to a 1913 census, and the pig population rose from 3.8 to 25 million during this period.[34] Germany was not exceptional in this global trend of expanding domestication in the first half of the twentieth century. Yet the loss in wildlife and the rise in domestic animals pertains to the relevance of the romanticization of Germanic animals and extinction in the Third Reich, and many Germans at the time would have encountered most animals only in a domesticated context on farms or in zoos.

While wild animal numbers were plunging, the Nazi party rallied around the idealization of certain animals as purebred or of robust stock in a time when many animals in Germany had been domesticated with controlled breeding programs or semi-domesticated through animal management efforts. Even animals vaunted as wild, such as red deer and chamois, were rigorously monitored by Reich government agencies supervising hunting. The zoologist Franz Graf Zedwitz notes in his 1937 book *German Wildlife*, "The hunt on our deer has gradually changed from a pure shooting hunt to a well thought-through game conservation. The laws of the Third Reich make sure that the wild does not get out of control."[35] It is estimated that there were 679 animal protection societies in Germany in the early 1930s,[36] supporting a widespread belief that the welfare and fitness of the nation's animals had direct bearing on National Socialist aims of "biocracy." New laws passed in the Third Reich promoted both animal conservation and hunting, viewing these objectives not as contradictory but, rather, as contributing to the health, robustness, and better breeding of Germanic natures.

The heavily administrated and biopolitically regulated animal life in Germany comingled with deeply romanticized idealizations of Germanic nature as the primal spiritual sustenance of the Volk. Yet by the time of the Nazis, the countryside of Germany had a very limited range and a diminishing number of wild animals, such that the celebration of an authentically German nature coincided with a highly depleted and mastered landscape. The passing of the Reich Nature Protection Law of 1935 (*Reichsnaturschutzgesetz*) did establish a broad legislation aimed at species conservation and legalized the process of declaring regional nature reserves under a rubric of preservation that limited industrial development. The law, widely praised by German ecologists and landscape architects, also established four "National Nature Reserves," akin to national parks, that were rich in game and prized, in particular, by Hermann Göring, the *Reichsjägermeister* in charge of hunting policy.[37]

The overall aim of these nature reserve laws and wild animal protections, however, was to subsume animal life in keeping with the Reich's goal to solidify attachment to the National Socialist vision of devotion to the homeland. As Thomas Lekan points out, "The organismic language of community ecology thus served as a scientific justification for Blood and Soil in the Third Reich, transforming nature parks into outdoor laboratories for investigating the optimal environmental conditions for the Germanic race."[38]

This is the context in which Martin Heidegger issued his philosophical statements rhapsodizing pastoral nature during and just after the Third Reich—for example, his philosophy-poem, "When the cowbells keep tinkling from / the slopes of the mountain valley / where herds wander slowly . . ."[39] that opens the volume *Poetry, Language, Thought*. Heidegger also excoriated what he called treating the environment as "standing reserve" or "enframed" as a resource exploited for anthropocentric satisfaction. However, such pastoral preserves in Germany were themselves already enframed, and animals were treated largely as "standing reserve" within the context of practices of domestication, agriculture, forestry, and the setting aside of natural reserves, which were organized nationwide in keeping with a National-Socialist agenda.

Heidegger also declared that all animals are ontologically "poor in world"[40] and "captivated" (259) by their immediate surroundings—be they wild or domestic, from tiny insects to the blue whale—because they lack a capacity for reflectiveness or any apprehensiveness of metaphysics. Heidegger treated animals as an entire single category already definitionally fixed; he refused to delve philosophically into the variations among species, deeming such examination unnecessary since, in his view, no animals had fundamental access to language or metacognition. Animals were reified and stuck in ontological place—effectively enframed at a metaphysical level—whereas humans were fundamentally definitionally open and questioning. Heidegger found animals "poor in world" in the context of a world that already was poorer in animals in terms of diminished wildlife. The sense of animals as "captivated" by their limited surroundings, thus, overlaps with the increasingly captive practices of landscape management toward nationalist supremacy. Heidegger's refusal to open up the question of the animal, to consider how animals might question the ontological difference between humans and animals, reinforced the widespread sense of animals as fundamentally captivated and captured.

Although wild mammals in Europe during the first half of the twentieth century had significantly dwindled in population, very few were to become completely extinct. In fact, stretching to the end of the Pleistocene (the last ice age, circa 11,700 years ago), the only mammals to have gone extinct in Europe are the mammoth, the woolly rhinoceros, the Irish elk, the wild horse, and the aurochs (many other mammals have been extirpated from Europe but survive on other continents).

The story of how the aurochs caught the attention of a group of elites in the Third Reich stands out as a paradigmatic moment of the multiple, contradictory, and deeply politicized concepts of extinction promulgated by the Nazis. The aurochs (*Bos primigenius*), a large bovine and the progenitor of domestic cattle today, dwelled across the grasslands, lowland forests, and plains stretching from Europe to North Africa, into much of Asia, and through the Indian subcontinent. The animal was domesticated about 10,000 years ago, a process which eventually led to the common cow (*Bos taurus*). The wild aurochs dwindled in number due to hunting until only small populations remained in Eastern Europe by the thirteenth century, where hunting of the animal was restricted to an elite class of nobles. The last known aurochs died in the Jaktorów Forest in Poland, on the king's hunting reserve, in 1627. Roaming across wide swaths of Europe, Asia, and North Africa, the aurochs had no specific Germanic identity, although the name "aurochs" or *ur-ochse* is of Germanic origin.

The possibility of reanimating the animal, and doing so in the name of the Reich, became a cause célèbre for the zoologist brothers Lutz and Heinz Heck, who saw in the aurochs a means by which to test their theories on reviving extinct animals by a process of back-breeding. The Heck brothers claimed that a breeder could select remnant aurochs traits from the races of existing cattle through controlled breeding methods to re-create the progenitor of the race. Lutz Heck, who served as the director of the Berlin Zoo during the Third Reich, and Heinz Heck, director of the Munich Zoo, each combed Europe for subspecies of cattle they thought had traits that harbored evidence of aurochs lineage. The results of their work are now called "Heck cattle," and researchers today do not identify the animal as having the genotype or phenotype of the aurochs.

Lutz Heck cuts a remarkable, outsized figure, part scientist, part opportunist, who would go on archaic-style hunting expeditions with Hermann Göring. He helped build the Berlin Zoo into one of the largest collections of animals in the world during his tenure as director, in part by pilfering animals from other zoos, most notably from the Warsaw Zoo, that had an abundance of rare animals Heck coveted, a story told in Diane Ackerman's *The Zookeeper's Wife*. Heck also traveled extensively in Canada in the 1920s, full of admiration for the landscape and its animals. He helped bring Canadian bison to Germany to breed with the few remaining woodland bison, in an attempt to boost the hardiness of the herd and spur further population growth (of course, disregarding the genetic differences or "purity" of the races).

Heck, who began the project to revive the aurochs with his brother in the mid-1920s, viewed de-extinction as a biological tool, along with de-domestication, that would contribute to the recovery of a romantic Germanic landscape. Although his zoo was situated in the highly urban site of Berlin and featured semi-domesticated animals in enclosed settings,

Heck disparaged city life and viewed domesticated animals as deprived of vitality and contact with wild nature. Heck viewed the aurochs as by definition preceding domestication, and he gravitated to the animal in particular for its mythological reputation as a fierce creature that could be fallen only by a worthy hunter. Heck was fond of the mystique of the aurochs and associated the animal with figures of the bull god in Egyptian and Greek mythology. The distinct Germanic legacy for the animal, according to Heck, could be found in a passage in the *Nibelungenlied* detailing a hunt by Siegfried: "Now Siegfried slew a bison and an elk, of wild aurochsen four, and a grim Schlech [wild stallion]."[41] Heck unabashedly wanted to recreate a Germany where one could hunt as Siegfried had done. To bring back the wild, for Heck, meant being able to hunt as these archaic and mythological Teutons had.

In his autobiography *Animals, My Adventure*, Heck discusses how his approach to reviving the aurochs centered on breeding these animals for what he perceived as the relevant physiological, aesthetic, and behavioral qualities, such as the color of its hide and the animal's "fighting spirit":

> The aurochs still lives on in some primitive races of cattle maintained for their yield or their fighting spirit, and its heritable constitution has largely remained untouched in these descendants. It is inaccurate, therefore, to say that the aurochs is extinct. No creature is extinct if the elements of its heritable constitution are still to be found in living descendants. All that needs to be done is to apply the experience of the breeder to the assembling of the inherited elements scattered among these descendants. To this end a sure eye is needed for the primitive qualities of this or that race of cattle, the nearest approach to the horn of the aurochs, or its combativeness, or its long legs, or the colour of its hide, or to the small udder of the wild animal (142).

As his guide to what the aurochs had looked like, Heck used imagery from Cro-Magnon-era cave paintings and drawings made during the Renaissance that were claimed to be direct depictions of the animal. He spent years picking and choosing among cattle, taking some from islands off Italy, and others from French and Spanish bullfighting breeders.

After over a decade of the application of back-breeding, Heck declared the aurochs had been revived. He also claimed successful de-extinction with another animal for which he had performed a similar operation, the tarpan (*Equus ferus ferus*). Basking in self-congratulations, Heck announced, "Perhaps, however, for me as a research zoologist the most wonderful and adventurous experience was the moment when, after years of painstaking experiment in breeding, carried out in Berlin and Munich after discussion with my brother Heinz . . . animals universally regarded as extinct reappeared. The breeding back of the aurochs and of the tarpan, the mouse-grey wild horse of ages past, had been accomplished!" (v). With Göring's permission, Heck reintroduced the ersatz aurochs, along with other large mammalian game, into the Białowieża Forest in Poland. Lutz Heck's cattle did not survive the war, but a small population of cattle bred by Heinz Heck lasted in Munich, and descendants of this group survive still today.

In his autobiography, Heck recites that he received research support and drew his ideas from "Professor [Erwin] Baur, the eminent plant breeder" and Eugen Fischer, "the first to investigate hereditary behaviour in accordance with the Mendelian rules in man" (143). Baur and Fischer, along with Fritz Lenz, wrote *Foundations of Human Heredity Teaching and Racial Hygiene*, published first in 1921 and appearing in several editions into the 1940s. The book served as an enormous inspiration to Hitler in developing his ideas on race, and in the time of the Third Reich, it was acclaimed as one of the main scientific standards for theory and practice of eugenics, including sterilization.[42]

After mentioning these scientists—in a book written well after the war and published in 1952—Heck adds, "The underlying idea of the plan for the re-breeding of the aurochs was quite clear, and its scientific basis was unquestionable" (143). Heck directly refers to the discipline of racial heredity and notions of eugenic control over life as his guide, seeing in the de-extinction of the aurochs the restoration of the bovine's more vigorous and authentic primitive bloodline. By using racial science for purposes of de-extinction, Heck hoped to prove the legitimacy and usefulness of both disciplines. Yet the case could be made that de-extinction via backbreeding actually contravened one of the central tenets of National Socialist racial biology, namely that racial characteristics were fixed by both blood and soil, and that racial types among humans and nonhuman animals would always exhibit certain distinguishing tendencies. Backbreeding showed evidence of the plasticity of racial markers and the historical malleability, rather than purity or unchanging authenticity, of natural-cultural characteristics. De-extinction demonstrated that speciation and extinction could be viewed as complex processes that were not linear or fated.

Jamie Lorimer and Clemens Driessen, in their study of the breeding practices of the Hecks, indicate that "the brothers seemed unconcerned with genetic health and purity"[43] in their combinations of different races of cattle. Lorimer and Driessen add that the Heck brothers presumed "that the vigorous wild traits of the aurochs would predominate in any breeding activity—contrary to eugenicist anxieties about the weak outbreeding the strong—and thus restoration by recombination could be easily achieved" (637). The Heck brothers' breeding methods, relying on Mendelian heredity models that predate knowledge of DNA, did not recreate the genome of the aurochs; rather, they created a more robust version of the domestic cow that managed to survive in the wild with a genome that emphasized a certain horn shape, fur color, and larger body type.

Lutz Heck wrote explicitly of his aims to use back-breeding methods to achieve de-domestication as well as de-extinction to accomplish the restoration of an authentic Germanic nature. Both extinction and de-extinction would contribute to the immunitary project of Nazism. Encountering and controlling these extremes of life and death supposedly would have a direct effect on the health and salvational horizon of the Aryan body in its Germanic natural surrounds. The project of de-extinction imbricated the redemptionist biology of species revival with the rhetoric of racial and national dominance. Restoring the aurochs contributed to a broader attempt to redefine the phenomena of extinction as contributing to German nation-building in which the return of lost or depleted animals could be taken as central to the revival of Aryan-ness. The return of the aurochs would herald a return of an aristocratic and mythic age of the Teutons, which could be remade with modern conveniences and the latest scientific knowledge.

The work of the Heck brothers shows the degree to which animal science and mythic attachments to an Aryan sense of "animality" in the Third Reich mattered in shaping the meaning of extinction and contributing to the definition of Germanic animals as having redemptive associations, while non-Germanic humans could be extinctable in advance. There is no other historical case of genocide in which de-extinction played such a significant role in contributing to the oppressive biopolitical regime. The Nazi regime sought biopolitical power over all of life and death, seeking control over extinction in both directions, such that species could be made and unmade at will. Acts of murder contributed to the same motive to Germanicize the landscape as acts of de-extinction.

While the Heck brothers' efforts to bring back the aurochs was not considered successful, their work inspired other scientists to take up the effort in recent decades. Since 2008, renewed work to revive the aurochs has been led by the Tauros Programme, run by the Tauros Foundation in the Netherlands and the pan-European Rewilding Europe initiative. The group has compiled a database of the full genome of some of the aurochsen races, and is using the genomic information to identify similar gene patterns in existing cattle breeds. The Tauros Programme will use breeding methods similar to those used by the Hecks to achieve what they call "aurochs 2.0," or the new animal "tauros" (the Greek word for bull).[44] As the ensuing breeds take on more characteristics, the remade aurochs/tauros will be released into several sites across Europe as part of a larger project to boost the wild animal population across the continent.

The scientists involved in the project, like the Heck brothers, are keenly aware of the cultural and mythological resonance of the animal. In the publication *The Aurochs: Born to Be Wild*, which describes their work, the organizers of the Tauros Programme point to the Greek myth of Europa being taken captive on the back of a bull. The publication integrates this story with the claim that the aurochs and its descendants were central in establishing European civilization. "The aurochs has always been at the very root of the whole idea of a continent called Europe. It is in fact our continent's defining animal."[45] According to the framing of this de-extinction project, if the domestication of the aurochs helped launch European civilization, the revival of the aurochs will help launch a new Europe, embracing both urban and wild pasts and futures.

However, there is significant debate today on the bioconservational value of de-extinction. In a time when the overall rate of animal extinctions is increasing, some argue it is a distraction to spend time and money trying to bring back lost species. Others argue that such science brings much needed attention overall to extinctions and may be a scalable science in the future to stem the rising extinction rate. I address in further detail the ecological and ethical issues of de-extinction in chapter six, but it should be noted that reviving just one iconic animal, such as the aurochs, would be only a minor contribution to restoring Europe's biodiversity. As with the Heck brothers, the rhetoric of conservation employed by the Tauros Programme is made for a project that is as much about an investment in a new era of European biopolitics as it is about the interest of the animal.

While the breeders of the aurochs claim that reviving the animal will lead to a broader effort to preserve the grasslands of Europe, there is no urgent ecological demand or biological niche that needs filling by this animal. This does not mean it is inconsequential to revive the aurochs, only that de-extinction is an optional rather than a necessary contribution to conservation or re-wilding. It is important to provide a critical history of the interest in the de-extinction of an animal, which can have a prominent symbolic and biopolitical importance—and would certainly contribute to another redefinition of extinction—that should be understood in connection with the ecological claims made in its name.

## Biopolitical Theory and Extinction

Hannah Arendt, in *On the Origins of Totalitarianism*, claimed that the extermination camps marked a political turning point that exposed how the natural rights of humans dissolved just at the very moment they were most needed. Under the Third Reich, as Jews became more persecuted and dispossessed of citizenship and basic human dignities, the human rights that were argued to be inalienable proved to be powerless without institutional support or legal enforcement on a national or international level. As Arendt observed, "It seems that a man who is nothing but a man has lost the very qualities which make it possible for other people to treat him as a fellow-man."[46] Arendt compellingly discerned the paradox that to experience oneself as merely of the human species, bereft of political and social standing, is the moment when one is least human; that is, on the brink of losing one's human existence and already being cast into an inhuman condition. To be merely human is not to rediscover the origins of human nature but to no longer be within the ambit of the human condition.

Arendt did not think any philosophical or political definition of the human could exclude the threat of enforced extinction perpetrated by human groups upon each other. She believed there could be no everlasting remedy to this danger, and no political axioms or definitions of the human on their own could prevent a situation in which humans could revoke other humans' entire juridical and political standing. Human rights, Arendt insisted, could be realized only inside democratic political and juridical institutions—in which all are granted equal standing in principle—that must be built and maintained across generations with strenuous vigilance and under the umbrella of nations and international courts that can defend these institutions and principles with force.

Arendt identified as most dangerous any political form that took its defining principles from biology, race, or some other non-democratic essence identified in nature and claimed to be embedded in the living body. In the Third Reich, the category of the individual, with its autonomous status, dignities, and responsibilities, was discarded in declaring persons either biologically superior or biologically inferior. For Arendt, it will always be fatal to democracy to incorporate any biological element into politics because, in her thinking, the political must be the realm of the ability to make free deliberations and actions, while biology is the realm of necessity and physical attributes that are given rather than made. As a result, Arendt construes democratic politics as the realm that divides the human from the biological and the animal; this type of politics is the counter to a politics of "life" that instrumentalizes the political in service of something inherently not available to democratic deliberation such as natural hierarchy or physical survival. However, as we will see, this position prevents Arendt from articulating an ecological politics counter to the Nazi effort to enact a political "applied biology."

Despite Arendt's aspirations that political actions oriented toward building durable democratic institutions should be considered separate from the work of maintaining biological life, political decisions are made every day on the life and death of human and nonhuman animals according to a range of social, biological, and medical categorizations. Arendt hoped a clear distinction between the free political activity of humanity and the compulsory, subsistence-level of animal existence of humans would secure the foundations of liberal politics. As a consequence, she could not think about the biopolitical foundations of politics as well as the political implications of interconnected and interdependent species conditions.

In her assessment of Adolf Eichmann's trial, she asserted that Eichmann could be put to death for his complicity in mass murder, adding that Eichmann's refusal to share the Earth constituted part of his crime. In a concluding statement written in the persona of the sentencing judge, Arendt crafts her own reasoning for why not sharing the Earth justifies a death sentence for Eichmann. "Just as you supported and carried out a policy of not wanting to share the earth with the Jewish people and the people of a number of other nations—as though you and your superiors had any right to determine who should and who should not inhabit the world—we find that no one, that is, no member of the human race, can be expected to want to share the earth with you. This is the reason, and the only reason, you must hang."[47]

Determining who should share the Earth is synonymous with the doctrine of extinctionism perpetrated by the Nazis. However, Arendt's reference to sharing the Earth requires a biological and ecological sensibility that is incorporated into political laws and ethical practices to fully understand what sharing the planet entails. If "to share the earth" will become a law, it will involve the imbrication of biological and legal conditions (without collapsing these terms into each other), which Arendt had earlier denied.

In Giorgio Agamben's writings explicating the biopolitics of Nazism, he discusses Arendt's insistence on thinking of democratic political theory as distinct from the politicization of biological life and death (Arendt is clearly influenced by existential thinkers who distinguished between the purported fixity and determinism of the biological body and the experiential, malleable, and indeterminate human condition). In Agamben's view, further democratic human rights laws and institutional support for upholding these laws will continue to falter at extreme moments because they do not examine how the institution of law itself is founded on a logic that includes some forms of life while fundamentally excluding others, thus creating the category of biological abandonment. Those who are included within the law are treated as sovereign subjects of equal standing, while those excluded potentially are treated as what Agamben calls "bare life." Bare life is mere biological life that is shorn of norms and value, where "morality and humanity themselves are called into question."[48]

Agamben argues that, in modern biopolitics, all life is politicized through these schemas of sovereignty and bare life that collapse the existential distinction between the brute facts of physical life/death and the interpretation of lived phenomena as "conditions" that are the remit of political morality. Sovereign power over determinations of life and death, even if diffused through medical and economic as well as political fields, dissolve this fact/value distinction while tending toward extreme powers of judgment on the worthiness or unworthiness of "bare" existence. Thus, "the fundamental biopolitical signature of modernity [is] the decision on the value (or nonvalue) of life as such."[49]

Bare life is not the same as the mark of mere animal or species existence, since it is a politically produced condition. "Not simple natural life, but life exposed to death (bare life or sacred life) is the originary political element" (88). Agamben claims the Holocaust reveals one of the essential truths of modern biopolitics whereby all of existence (human or animal) is so thoroughly politicized to the extent that all of life is exposed to sovereign decisions over value or expendability and is, potentially, treatable as "bare life." Even if more laws are declared to protect human rights or animal rights, these do not fundamentally change the foundations of national, legal, and political doctrines that continue to function by requiring life to pass through a biopolitical decision that claims the privilege to determine which bodies do or do not have worthiness or legal standing. Life that is determined to be "bare life" will be deemed expendable, sacrificable, and, ultimately, extinctable if situations call for it. An entire species can be reduced to bare life, such as the bison discussed in chapter one.

In *Remnants of Auschwitz*, Agamben argues for a new kind of juridical basis for testimony situated at the level of those who have become "bare life" and are refused conventional political and legal forums of redress. The inability to testify, yet to still be a witness and to remain a human being regardless of the loss of one's humanity, becomes the

paradoxical foundation for a new political ontology that Agamben inti-
mates will bypass the reproduction of bare life. Following from Primo
Levi's account of the camps, Agamben insists a new ontology and ethics
must emerge from the realization in the Holocaust that "human beings
are human insofar as they are not human."[50] He also adds, "human beings are human insofar as they bear witness to the inhuman" (121). These
are paradoxical formulations that insist that a new sense of the commonality of human beings must be reconstructed from the point at which
human and inhuman "are coextensive and, at the same time, non-
coincident" (151). Humans are "not human" because no set category,
predicate, humanist principles, or law secures the humanity of humans
or protects from their inhuman treatment. What Agamben is pointing to
is how, even when one loses the capacity to say one is human, even where
speech becomes impossible, one still remains human because this condition testifies to a lack in self-capacity and a lacuna in language that
traverses both humanity and inhumanity. It is this non-coincidence (between potentiality and capacity, or testimony and language) that both
leaves a gap between humans and any substantive universalist claims for
defining the human and also ties humans together in a common, precarious existence.

This "original disjunction" (123) evident in the structures of language
and experience cannot be appropriated and already is implicated in the
overlap/non-coincidence of the human and the inhuman. There is a collective impossibility of appropriating humanity (or really any signifier)—
of totalizing the human into some form of transcendental fullness, but
also of totally denying someone inclusion in humanity by trying to
claim a partiality as full humanity—from which humans cannot turn
away, no matter how violently they destroy each other. There is an incapacity within the human to totalize or fully appropriate the human and
an incapacity to fully evade or be dissociated from the human. Since
this lack or potentiality—a non-essence, an unsayable element within
what is sayable concerning the human, and an open possibility that
traverses both the human and the inhuman—itself cannot be assumed or
destroyed, "the human being is what remains after the destruction of the
human being" (134).

This new testimonial principle seems very meager, but one of the
powerful effects of this claim is to nullify all biopolitical discourses that
try to break up, partialize, or exclude humans from the category of humanity. Humans will remain human not because of their designation of
belonging to the abstract name of humanity (which can be resignified and
brutally revoked from some humans) but because of the tautological fact

CONCEPTS OF EXTINCTION IN THE HOLOCAUST / 157

that to be human is to be human even when one is forced to bear witness to the loss of one's humanness. The claim "human beings are human insofar as they are not human" serves only to refuse the premise that humans have to qualify or nominate themselves as human in the first place. Most importantly for Agamben, the paradoxical definition of being human via a shared impossibility admits no reference to biological qualities or species specifications, since to admit any of these criteria in the qualifications of being human is to condemn humanity to a history of biopolitical decisions.

Agamben's minimal formulation still needs to be implemented and cared for in everyday juridical conditions, as Arendt would have insisted. It is not clear if Agamben foresees the realization of any laws created from such stark axioms, since he also claims such minimal testimony to be prior to sovereign and juridical power. At the zero point of bare life—in an extinction event—what do terms like testimony, life, and language mean? Agamben continues to use these terms, but do they still function when pushed to their existential limits? Would we not need institutional and juridical understanding of these terms to still be able to recognize them in extreme conditions?

Beyond the problem of Agamben's reference to a pre-juridical realm and "form of life" separate from the institution of the law and biopolitics, there also is the issue of Agamben's evacuating definitions of the human from not just biological references but ecological ones, as well. Agamben's theorizations of nature are primarily textual readings from philosophy and academic biology and lack contextualization with the status of animals and environments during the Reich. His articulation of the human "beyond any figure of relation,"[51] including biological terms that Agamben perceives as doomed by an organicist essentialism that is captured by sovereign power's inclusion of life into politics, is posited as a means to extricate existence from biopolitical hierarchies. Agamben is rightly wary against discourses of disembodiment, and in many other texts has sought to establish a philosophical claim to sensual embodiment that would be impervious to biopolitical decisionism. However, his evocation of an extreme "remainder" also risks removing humanity from ecological relationships and interspecies connections that are not reducible to contestations over biopolitical norms (and even though Agamben elsewhere has written on animals, he has done so as way to trace the history of biopolitics and to affirm a sort of idealized post-juridical and post-ecological realm of redeemed human-animal community[52]). Agamben is so focused on bare life and redeemed life that he is unable to attend to an everyday ecological life that is neither doomed nor salvationist. Agamben misses how even the "remnant" is ecologically situated and entangled in natural-cultural exchanges that are not reducible to biological essentialisms.

A similar problem appears in the work of Roberto Esposito, who follows Agamben in analyzing how the extremes of biopolitics in the Third Reich became exemplary for modern political theory. Esposito argues that Nazism pursued an immunitary logic to its extreme limit by forcing the politics of protecting and securing life to converge with a systematic incorporation and control over death (thanatopolitics) as the very means by which to foster such life.[53] Esposito traces how biopolitics transformed into thanatopolitics through incorporating immunitary practices in medicine, eugenics, and hygiene into the juridical sphere and treated as tying the organic to the nationalistic. These immunitary programs claimed to protect the biological body/nation by everywhere attacking any life deemed not pure life. This convergence of biopolitics and thanatopolitics in the Third Reich culminated to the point that Nazism no longer distinguished between life and death by treating any and every life as killable in advance.

Such thanatopolitics was first addressed toward all non-Germans, and eventually directed onto the German national body by the end of the war. Both the biopolitical and thanatopolitical frames help explain how Nazism sought to synthesize its own definition of extinction with everyday political and juridical power over life and death. Esposito shifts at this point in his work to propose a politics of life and embodiment going forward that would forestall the collapse of biopolitics into thanatopolitics. He argues for a counter-tradition of biopolitics that would affirm the plurality of bodies and persons across diffuse and inclusive biological norms. However, in Esposito's own writings, this argument for a nonessentialist, non-pure, non-unified pluralistic body as the condition for a more just politics requires engaging with animals and environments as more than generic additions to a human-centric pluralism. Furthermore, it requires thinking about how biological pluralism itself is entangled with natural and cultural histories of extinction.

The Nazi plan to facilitate "organized extinction," which dovetailed with their interest in subsuming all of life and the environment into a total plan for German *Lebensraum*, revealed another kind of testimonial remainder that can be described as ecological because this experience is constituted only in terms of environmental and species relationships that are not reducible to any anthropocentrism, be it of Germanic or pan-humanist origin. In writings by survivors of World War II concentration camps, references to the undoing of humanity could appear side-by-side with recognitions that the immediate ecological context of the camps—the woods, the soil, the clouds, the sky—were not themselves abolished or redefined into complicity with extinction by the Nazis. The ecological consciousness of survivor narratives stands as a crucial reclamation of the environmental contexts in which the Nazis tried but failed to redefine extinction under their own terms. This ecological condition, even in the midst of extreme oppression, did not become wholly subsumable to Nazi state biopower.

Further insight into how the extinctionary agenda of the Nazis stood in contrast to the immediate ecological relationships of species appears in a powerful passage in Robert Antelme's survivor account, *The Human Race* (*L'Espèce humaine*), published in 1947. Antelme, who was not Jewish but was a French resistor captured by the SS in the summer of 1944, tells of his internment at Buchenwald and Gandersheim. He is sent on a death march to Dachau just at the close of the war. Antelme uses much of the same imagery as Primo Levi (they published their witness accounts at about the same time) in explicating how the camps forced a reckoning with what happens when everyone is treated like a bare biological entity without recourse to the dignities granted to being human. In an extended passage, Antelme writes of his coming to understand that the human species still meant something ecologically and politically even after the Nazi attempt to destroy the species unity of humanity. Antelme's reflections occur while he is forced on a group death march to pass through valleys and forests in the early hours as dawn emerges, and he notices the trees, animals, and insects surrounding the group.

> To us who look so like animals any animal has taken on qualities of magnificence; to us who are so similar to any rotting plant, that plant's destiny seems as luxurious as a destiny that concludes with dying in bed. We have come to resemble whatever fights simply to eat, and dies from not eating; come to where we exist on the level of some other species, which will never be ours and towards which we are tending. But this other species which at least lives according to its own authentic law—animals cannot become more animal-like—appears to us as magnificent as our "true" species, whose law may also be to lead us here to where we are. Yet there is no ambiguity: we're still men, and we shall not end otherwise than as men. The distance separating us from another species is still intact. It is not historical. It's an SS fantasy to believe that we have an historical mission to change species, and as this mutation is occurring too slowly, they kill. No, this extraordinary sickness is nothing other

than a culminating moment in man's history. And that means two things. First, that the solidity and stability of the species is being put to the test. Next, that the variety of the relationships between men, their color, their customs, the classes they are formed into mask a truth that here, at the boundary of nature, at the point where we approach our limits, appears with absolute clarity: namely, that there are not several human races [espèces], there is only one human race [espèce]. It's because we're men like them that the SS will finally prove powerless before us. It's because they shall have sought to call the unity of this human race [espèce] into question that they'll be finally crushed. Yet their behavior, and our situation, are only a magnification, an extreme caricature—in which nobody wants or is perhaps able to recognize himself—of forms of behavior and of situations that exist in the world, that even make up the existence of that older "real world" we dream about. For in fact everything happens in that world as though there were a number of human species, or, rather, as though belonging to a single human species wasn't certain, as though you could join the species or leave it, could be halfway in it or belong to it fully, or never belong to it, try though you might for generations, division into races or classes being the canon of the species and sustaining the axiom we're always prepared to use, the ultimate line of defense: "They aren't people like us."

And so, seen from here, luxuriousness is the property of the animal, and divineness is the property of trees, and we are unable to become either animals or trees. We are not able to, and the SS cannot make us succeed in it. And it is just when it has taken on the most hideous shape, it is just when it is about to become our own face— that is when the mask falls. And if, at that moment, we believe what, here, is certainly that which requires the most considerable effort to believe, that "The SS are only men like ourselves"; if, at the mo- ment when the distance between beings is at its greatest, at the moment when the subjugation of some and the power of others have attained such limits as to seem frozen into some supernatural distinction; if, facing nature, or facing death, we can perceive no substantial difference between the SS and ourselves, then we have to say that there is only one human race [espèce]. And we have to say that everything in the world that masks this unity, everything that places beings in situations of exploitation and subjugation and thereby implies the existence of various species of mankind, is false and mad; and that we have proof of this here, the most irrefutable proof, since the worst of victims cannot do otherwise than establish that, in its worst exercise, the executioner's power cannot be other than one of the powers that men have, the power of murder. He can kill a man, but he can't change him into something else.[54]

In this long passage, Antelme writes of how in the concentration camp the ruination of the human brings out several meanings of the term *species*. Three different meanings can be highlighted. First, the human as species is revealed at points of extreme degradation, suffering, and death to be just a biological being among others, "something they fill with water and pisses a lot" (95). In this case, one is first a mere biological existence ("*espèce*"), and then sometimes treated as a human or not a human. A second meaning of species is that there is a kind of acceptance of species as being and acting themselves by other animals. The lives of animals appear "magnificent" and "luxurious" in going about their daily being, and they do not try to make other species not be what they are. Any given animal will see humans as humans, flies as flies, and dogs as dogs, who live and die according to the same laws as every other animal. This is not an essentialist biological statement but, rather, a form of shared acknowledgment and relationship. Animals do not question the very animality of each other—they do not try to change a species into something else. A third sense of species is in the claim for a minimal solidarity among humans due to the inherent resistance of any species to be deprived of its speciesness even upon death. At the limit of the human, one doesn't become literally another animal—even when one human kills another, the victim dies as a human being, not as something else; for example, as a vermin or a louse. Living or dying as a member of one's species is an ecological reality that remained as such even in Nazism.[55]

In the camps, Antelme perceives human history and natural history converge biopolitically in a place where "every possible degree of oppression existed" (4), but he finds that certain ecological realities of natural history did not go away. The surrounding trees and animals testify to an environs that is not just a world of human history, where humans can redefine everything concerning biological and physical reality according to the whims of the dominant power. The ecological and species history of being human comes to confront the biopolitical redefinition of the human at these extreme moments and does not yield or disappear even in death. The "magnificence" of animals and plants cannot be dissolved or fully appropriated by whatever agents in human history would claim that trees are not trees and humans are not humans.

To be sure, all these species and ecologies can be annihilated. Going even further than just killing people, the Nazis sought to redefine the parameters of the human and, effectively, every other species. Antelme's restating that the human species still remains one—that the species definition of the human has not changed—constitutes an act of resistance even when all the dignities of being human have been destroyed. To revisit Agamben's articulation of the remainder as witness, humans remain human even after being expelled from the category of humanity because they are not thereby removed from being a species with a human body among other living beings. Even the expelled are included within the midst of the lives of other animals as present together in an ecological and embodied-relational sense, together exposed to the sky, dirt, wind, and rain. Perhaps, then, the better term is not remnant but relation. Antelme concludes that there is a minimal dignity in being a species on Earth even when there is no dignity among humans, because the ecological condition still persists and humans still cohabit and are co-defined with other plants and animals. Such collaborative definitions of terms like species and extinction already serve to refute anthropocentrism and the nefarious will to appropriate everything or redefine life according to dominant human powers.

## Toward a New Universal Law after the Holocaust

In the Third Reich, practices of extinction toward undesirable non-Germans became promoted as an obligation and a duty to immunize the German state-body. After World War II, in addition to the new universal law that there never be another Holocaust, another law can be added: there can be no declaration that a life form is extinctable in advance. Extinction can never be a right legislated metaphysically in advance to annihilate one species in the name of another species. However, this is a complicated new universal law, for it also is impossible to say that there should be no more extinctions at all.

There is no form of biological life that can avoid extinction, and there is no form of ecology in which extinction is rendered permanently impossible. No animal should be condemned to extinction in advance, but extinctions can and will still happen. Controversially, it also can be the case that an extinction may be a justifiable action. It is possible to make the case that extinction for medical reasons is permitted—the eradication of viruses, bacterial plagues, and disease-carrying mosquitos is justified to alleviate human suffering. Such extinctions are medically legitimate on a case-by-case basis but cannot be justified by appeal to metaphysically predetermined principles on a valuation of life that follows from a refusal to share the Earth. It is possible to maintain the doctrine that all species have equal physical claim to sharing the Earth while also up-holding the ethically pragmatic principle that human suffering ought to be mitigated. Even medical justification for eradication of a life form must remain always questionable, never established as an unquestioned metaphysical axiom. Every extinction event must be questioned before, during, and after the case.

There is never an obligation to extinction; it can only ever be a difficult choice that will involve case-by-case compromises and cautious judgments as to the best form of care and reduction of suffering, not a new universal law. One can make a universal law that no life form is extinctable in advance; however, one cannot make it a new universal law that there never be extinctions, because such a law, were it possible, would change all life and transform all ecology into something else (this problem returns with regard to contemporary biotechnology advancements toward the feasibility of de-extinction and is discussed further in chapter six). Humans must do all they can to be vigilant to prevent or avoid all extinctions in the present. Humans also have a duty to not be the direct or indirect causes of species extinction due to anthropogenic climate change, unsustainable hunting and fishing, habitat destruction, or pollution. Humans should not be the deciders of which species get to share the Earth—except in rare cases where extinction has a medical basis to alleviate great suffering. But even in such cases, the medical justification for eradicating a species such as a virus should not become a metaphysical principle that justifies an extinction in advance. Another way of saying this is that humans should not be granted the power in advance to permanently redefine extinction as no longer an evolutionary and ecological process but, instead, strictly a biopolitical judgment from now on.

There is an additional obligation implicit in the duty to never turn extinction into a metaphysical determination made in advance: there must be the utmost vigilance over the terms, concepts, and rhetoric of extinction and genocide. One characteristic of the extremification of modern biopolitics is the mobilization of the discourse of "existential crises," the claim, whether bolstered by evidence or metaphorical hyperbole, that some form of extinction is at stake for a particular group, and, thus, a particular group or nation is justified to some degree in taking radical measures in the name of existence itself. How will we assess when actual existential crises—when existence for all or the existential condition itself is at stake—are to be distinguished from other phenomena that use the rhetoric of existential threat?

Consider the current phenomena of the rise of the phrase "white genocide"—the claim made by white nationalists and neo-Nazis that white people and whiteness are mortally threatened due to demographic and cultural trends. Whatever whiteness means today, there is no organized dehumanization and persecution of whites and whiteness anywhere and everywhere. There also is no crisis of demography for whites; there are more white people living today than at any other time in history. Nazis committed genocide; neo-Nazis wish to claim themselves the victims of something akin to genocide. Such usage makes "genocide" a meaningless referent, which is partly the point, since it sows confusion about what genocide means. White genocide is an attempt at another "organized lie" concerning the concept of extinction compounded by a biopolitical lie that whiteness is in danger.

Against these attempts to change the definition of extinction and genocide, it remains imperative to insist on the distinctions between defining extinction as something all species eventually face, the human history of extinctionary acts committed against other species, and the human act of genocide. It also is necessary to reflect carefully on how these distinctions have been combined and confused during historical events over the past two centuries. Understanding these distinctions and the uses of these terms will remain difficult, and any future posing of the question "What is extinction?" will have an impact across all these terms. Every situation concerning extinction and the use of extinction rhetoric requires a constant vigilance, but there is a lesson from Nazism that one should guard against becoming so obsessed about extinction that it becomes the dominant and ubiquitous frame for all events.

The Nazis made extinction all too present and too meaningful and determinative as the rationale for all relationships and actions no matter how mundane or regulatory. Reckoning with extinction today requires the utmost dedication to cultivating ways to share the benefits and burdens of being on Earth while maintaining the utmost attentiveness over any attempts to redefine extinction from a collective ecological responsibility into a biopolitical determination over what kind of lives are "worth living."

# PART III

# 5 / Critical Theory for the Critically Endangered

The global depletion of biodiversity over the last several decades has pushed many animal populations to precipitously low numbers. Some of these species will go extinct, some will recover to more sustainable population numbers, and some will dwell for a prolonged period in the uncertain, unsettled status of very small populations. The broader effects of the phenomena of extinction—the impact such biological extremes have on shaping the social, psychological, and cultural conditions of animal and human life and death—extends to animals that are not yet extinct but whose endangered status has already had transformative effects on the species' existential conditions. Alongside the rising numbers of lost animals, the category of critically endangered last animals has expanded rapidly in recent decades. The International Union for the Conservation of Nature defines "critically endangered" as a species that has lost greater than 80 to 90 percent of historical population numbers in the past decade.[1] Each species has a "minimum viable population threshold" needed to reproduce itself (for many animals, this means numbers between 250 and 500).[2] Even though biodiversity is increasingly calculated among other economic, nationalistic, ethical, and ecological "goods," conservation budgets and public interest in animal protection often is limited by competing interests. Concerted efforts to halt extinctions now frequently can result in animal species living out protracted lives in categories that the IUCN classifies as highly threatened.

As a result, there is an expanding category of minimal viable populations. Though such species may narrowly avoid extinction, they survive under its shadow. The lives of many last animals are now largely stuck in a strange holding pattern, where the animal is maintained in small numbers that are deemed minimally sustainable but not much more expandable. Living indefinitely at the existential edge, then, increasingly defines much of species life today.

What happens when a species declines toward "unviable" population numbers? In addition to the biological changes that affect species at low population numbers, what kinds of cultural and ontological changes happen to life lived in low numbers? The depletion of animal populations has become such a global phenomenon that, as J. B. MacKinnon puts it, "We live in a 10 Percent World."[3] To a striking degree, many animal populations worldwide have been diminished to numbers that total 10 percent or less of historical numbers. MacKinnon cites how many of the world's biggest fish populations, including tuna, cod, and sharks, have been reduced to below 10 percent of the levels in the recent past. A similar percentage of depletions applies to nearly all wild mammals across the globe, and most bird populations, as well, have seen precipitous declines.

Indeed, in many cases, 10 percent would be an optimistic number. Giant tortoises in the Galapagos declined from 250,000 in the 1500s to 15,000 today. Chimpanzees numbered 2 million as recently as 1900 but now number 150,000.[4] A stunning report issued by the World Wildlife Fund in 2018 calculated that up to 60 percent of wild vertebrate animals have been lost since the 1970s.[5] Not only population numbers but also the historical range of habitat for the vast majority of animals has been dramatically reduced. Tigers used to roam throughout nearly all of Asia and the Middle East, but now are found only in small pockets of territory, most notably in India, China, and Russia, less than 10 percent of their historical range. The wolf and grizzly bear used to range almost across all regions in Europe, Northern Asia, and North America; almost no bears remain in Europe, and the wolf has been eradicated in nearly all its former territory in Europe and much of North America.

These low populations and drastically diminished habitat ranges are the new norms for much of the non-human life on the planet. The terms of sustainability now refer to the struggle to even maintain these radically depleted numbers. In many cases, animals are allowed to thrive in protected zones but are practically powerless outside these parks and preserves. Indeed, it often is deemed a success when such animals survive in small but stable populations inside conservation enclaves. Sustainability for animals now mostly means existing in small, delimited, thinned out, and scrutinized habitats indefinitely. These enclaves are easier to visit because they are compact destinations and, therefore, support ecotourism and are more amenable to patrol against poachers.

Condensed biodiversity enclaves, which now largely describe the *de facto* habitats of the 10 percent, offer a special kind of animal capital. Inside these zones, economics and animal population size reach a kind of collusion. For example, lions used to roam almost all of Africa, the Middle East, Asia, and even much of Europe. Now that only approximately 40,000 lions remain, they can be kept in smaller, concentrated areas, which are easier to visit, manage, and keep separate from humans. A limited number of permits can be allowed for hunters willing to pay up to $30,000 (a recently quoted price to hunt a lion in Botswana), which provides the funds for further lion conservation and sustains local economies, which include the humans who face lion attacks from time to time and need a reason not to simply shoot the animals. If the lion population were larger, the permits would not be worth as much and there would be more attacks on humans, and, without the shadow of extinction looming, conservation groups would be less inclined to spend time and money on the animal's behalf.

Last animals receive huge investments in research, conservation, and public attention, but when numbers rebound, these resources often shift elsewhere according to a "triage" logic, which, in turn, tends to result in the animals returning to lower population numbers as conflicts with humans recommence. Paradoxically, fewer animal numbers may be the pathway to more biodiversity preservation in general.

Outside of these enclaves, there seems to be no reliable formula for human–animal coexistence for most of these last animals today, even though just as recently as a few generations ago, humans regularly encountered these animals in their range and abundance across the globe. Mackinnon, borrowing a term from psychology, describes this phenomenon as "shifting baseline syndrome" (16) because the depleted number and range of such last animals establishes the next generation's norm. Each generation, thereby, encounters fewer animals that occupy smaller ranges, and assumes that the current lay of the land represents the standard population size and habitat for the animals. Most people have become used to seeing a 10 percent world as nothing all that alarming—they have lost the memory of past abundances, having grown up only with vastly diminished animal populations.[6] A 10 percent world right now is very comfortable and convenient for most everyday routines. In many parts of the world, we see animals when we want to and keep them apart from us at all other times, for their safety and ours.

This new norm has come at a devastating cost to wild animal life. MacKinnon cites a phrase from the biologist Norman Myers, who called the rapid disappearance of animals in the past few centuries "the great dying" (35). Furthermore, nothing guarantees that animal populations will stabilize somehow at 10 percent, even in plush, well-demarcated parks and enclaves. Animal populations may continue to be depleted, as disruptions could be caused by other factors, including breakdowns in the food chain, global heating, disrupted migration corridors, and newly introduced predators and competitors. At such low population numbers, the cultural and ecological relationships animals foster with each other, and with humans, are continually placed under immense pressure and frequent disruption. Finally, it also might be the case that the genomes as well as the behavior of animals change at these low levels—some crucial gene variations may have been lost, and learned hunting, navigating, or reproducing skills may not be passed on consistently in smaller groups.[7] The previous historical range of the animal might already have been changed irreversibly into a different landscape; the animal may have fundamentally changed, as well.

This chapter presents case studies of three species—bison, tigers, and coral—each of which currently inhabit that 10 percent margin and whose lives have become intertwined with the recent history of extinction. I combine analysis of different forms of care and dilemmas of conservation specific to these three animals, while also turning to aesthetic representations of their lives to better understand what it is like for an animal to face precipitously low population numbers. These highly visible species play iconic roles in global attitudes toward extinction today. The permanent loss of any one of these emblematic species would reverberate so powerfully as to change the meanings of extinction. As last animals, they dwell in a gray zone where it is not clear whether they are in recovery or teetering on the precipice. They are suspended—though temporarily sustained—at the edge of extinction.

I have chosen to examine the current conditions of bison, tigers, and coral because each presents a distinct story of dramatic loss of population numbers, but, also, each has seen the very definition of its species existence intertwined with global conservation efforts motivated by multiple and often conflicting rationales. These "last animals" have come to serve as indexes for broader issues of the traumatic loss of biodiversity and population collapse in an era of globalization and global heating. They also have become animals invested with extraordinary hope of conservation science, multi-scalar community engagements, and concentrated efforts at connecting the restoration of fauna with social justice and anti-colonialism.

These species also benefit from the proliferation of cultural works dedicated to efforts to see the world from the animal's point of view. Understanding the situations of these animals requires a "multispecies studies" methodology, as articulated by Thom van Dooren, Eben Kirksey, and Ursula Münster, that involves multidisciplinary inquiry to achieve "careful and critical attention to the specificity of other species' lifeworlds."[8] Instead of theorizing "life itself" or "death itself" as concepts abstracted from ecological contexts and the actual lives of animals, an examination of the specific conditions of endangered animals requires understanding how definitions of life and death intertwine with wavering numbers of species and shifting norms of precarious life. Donna Haraway emphatically makes the point that it is important to cultivate ways of living and dying with animals in forms of being-with that avoid the polarizations of "awful or Edenic pasts and apocalyptic or salvific futures."[9] Rather than reproduce these categories of salvation or apocalypse, it is important to examine in detail the lives of last animals and to provide a critical theory for the critically endangered.

## How the Buffalo Roam Now

The condition of the plains bison today provides a striking example of how animal life is lived below 10 percent. While the nineteenth-century bison discussed in chapter one became a figure for extinction, the twentieth-century bison has become a figure for the biopolitics of repopulation after an extinction event. The recovery of the bison has had a broader cultural and scientific impact on thinking about the possible afterlives of animals in the wake of an extinction event.

Even as the bison were massacred in mass in the 1870s, a number of attempts to safeguard the species were launched by members of animal protection societies, concerned ranchers, Indigenous peoples living on the plains, and those who protested against the notion of rendering an animal extinct on principle. William Hornaday established the American Bison Society in 1905 to launch an organized preservation of the animal. Hornaday's group worked to track the remaining animals, manage a small herd, pass legislation that established no-hunting reserves, and develop research methods that would aid in growing the population back from such radically few numbers. The group achieved steady success in the first international attempt to use conservation science in conjunction with legislation to prevent extinction.[10] By the 1930s, the bison population had grown to over 25,000, in wild herds and some on ranches and in zoos, and the group declared their work accomplished and disbanded. Within a few decades, the population would continue to grow, plateauing to over 300,000 by 2000, but most of the population increase would be on private ranches, with the animal destined for food consumption.

Today, bison numbers still are well below 10 percent of historic populations—actually they are much closer to 1 percent, and of those remaining numbers, only about 10 percent live on land protected from hunting and ranching. But in recent decades, the bison, again, has caught the imagination of conservationists who desire to propose something big, bold, and captivating toward the idea of "rewilding," which involves reintroducing animals to past habitats to reconstruct fragmented and depleted ecosystems. Strong calls for various rewilding projects have been made by Gary Snyder, Ernest Callenbach, Dave Foreman, Paul Martin, Emma Marris, George Monbiot, and Marc Bekoff, among others.[11] Rewilding is not just about bringing the animals physically back; it also includes advocating for a spiritual and cultural reclamation associated with the return of specific regional species and ecosystems. While these rewilding propositions do give a nod to Indigenous efforts to connect Native resurgence with restoring the bison to the plains, they are not primarily driven by a decolonial impulse. These cases for rewilding the bison are motivated largely by a conviction that the new baseline of a 10 percent world is stultifying for both animal and human ecologies.

Bringing back the bison, then, for example, would be a way of returning to something essential about being human and animal that was lost in the era of high capitalism and declarations of "the end of nature." Further implied in rewilding is the notion that there would be an aesthetic reawakening of nature-cultures with the return of such animals, reconstituting a much-sought overlap of art and animal wildness. Snyder's poem "Home on the Range" formally enacts this aesthetic rewilding as it appropriates the nationalist nostalgia of the American folk song to envision a wholly bison-enabled way of life:

> Bison rumble-belly
> Bison shag coat
> Bison sniffing bison body
> Bison skull looking at the sweat lodge.
> Bison liver warm. Bison flea
> Bison paunch stew.
> Bison baby falls down.
> Bison skin home. Bison bedding,
> "Home on the Range."[12]

The actual presence of bison on the range may be severely diminished, but this poem rematerializes the missing animal by bringing the bison up close in the space of the poem and activating the sensuousness of both animal and reader. There is no subject, no "I" to mediate experience, just the directness of the bison that can be heard, touched, smelled, tasted, and everywhere seen. Such comprehensive rewilding suggests even a change in the power and purpose of our senses. Reader and bison seem to be in the same intimate space, as the poem does not distinguish between the "liver warm" inside the bison and the same object being consumed by a human. The only thing that indicates mediation in the poem are the quotation marks around the folk song "Home on the Range," a cue to the reader that this song and its title are being rewilded, too.

The intimacy of the speaker with the bison body is suggestive of the Plains Indigenous way of life, but the song is a settler tune. Re-wilding the imagination is an important step in dislocating the imagination from its settled ways, yet the transitions from textual rewilding to large-scale rewilding projects in developed landscapes—especially those with colonial histories—require close attention for the kinds of conservationist claims made on animality and sustainability together.

Ernest Callenbach, previously famed for *Ecotopia*, envisioned the return of the bison to the plains as a project that would integrate ecological and national renewal in his book *Bring Back the Buffalo! A Sustainable Future for America's Great Plains*.[13] Arguing for ecological sustainability as well as economic prosperity, Callenbach claimed the repopulation of the bison to be a convergence of idealist activism and pragmatic conservation. With the rewilding of the bison, Callenbach sought a post-utopian nationalist agenda rooted in agrarianism after his visions of a radical separatist ecotopia faded. Transitioning away from the ecotopian dream in the 1970s of pastoral abundance leading to social and sexual harmonies, this book calls for a future ecology to be focused on practical declarations of sustainability and economic boon managed by ranchers (primarily masculine) as well as state and corporate entities.

The steps to bison repopulation outlined by Callenbach retreat from demanding a whole new ecopolitical outlook. Callenbach's plan would be to coordinate government and private land purchasers to obtain a wide swath of the plains, land that is not arable without a tremendous input of fertilizers, petroleum-based machines, excessive aquifer usage, and financial subsidies. The huge revenues allotted by the U.S. government to farmers (whether or not they grow), instead, could be used to buy land back from farmers, presumably saving money in the long run and stemming the tendency of the government to prop up unprofitable farms. The purchases would be strategically adjacent, with the idea of creating corridors between national parks that follow the historic paths of bison migrations. Here "we must learn to 'think like bison'" in land use (29) so that bison and grassland develop with each other. Bison chew only the tops of native grasses, which allows them to regrow, while cattle, currently numbering over 110 million in the United States, rip grass from the roots (due to the cost of feeding them native grasses, cattle today are largely nourished on corn and other grains as substitutes). The return of the bison would lead, presumably, to widespread ecological restoration, with one species linking to a rewilding of many.

Callenbach projects that several million bison could be stocked on these lands, needing very little hands-on attention or fencing, as they live peacefully enough among humans if left alone. The multiple opportunities bison present to different stakeholders who rarely see eye-to-eye is a key benefit, according to Callenbach: bison could, variously, be publicly or privately owned, or under the auspices of Native American communities, or in nature conservancy preserves or large ranches. Here, Callenbach tries to piece together a version of sustainability in a contemporary economic parlance, as exemplified by Paul Hawken's *Natural Capitalism* (published in 1999), where private enterprise and environmental benefit are trumpeted as synergistic. Callenbach aims to see these animals as fully absorbed into capitalist systems of profit maximization and the industrial animal-rendering empire. Effectively, all the millions of bison would be available for rendering and consumption, and Callenbach envisions Indigenous and settler ranchers having free run of the killing of these animals as long as they do it to maximize food and keep the population of the herd restocked.

This declaration for sustainability on the plains is not a manifesto for animal rights or the flourishing of independent bison. The bison would return in huge numbers, but their increased population would be predicated on remaining as animal capital in a world of valorized ranchers, slaughterhouses, and meat retailers. Callenbach supplies many pages of commentary extolling how tasty the bison is, how healthy and low-fat the animal is for the hearty meat eater, and how promising the animal would be in the hands of fast-food operators. He suggests a "McBuff" burger (197) as the next great food item. His book is chock full of "bison entrepreneurs" (190), ranging from media mogul Ted Turner to small, one-person ranches, from big beef industries to mom-and-pop bison burger joints. Callenbach quotes Harold Danz, executive director of the National Bison Association, who unabashedly states: "To preserve buffalo, the best thing we can do is eat them. Animals that people eat do not become extinct" (186).

Lest this shock the sheepish animal welfare reader, Callenbach offers a quote from the trusted environmentalist poet Gary Snyder on the next page, who says with equanimity: "There is no death that is not somebody's food, no life that is not somebody's death. . . . Eating is a sacrament" (187). Along the lines of what Thom van Dooren calls "killing for conservation,"[14] here we have eating for conservation. Consuming and rendering is declared necessary to achieve rewilding. In the great American fantasy, eating more is the pathway to more abundance, while animal death is proposed as the plan to make animal numbers grow.

Callenbach hitches the notion of sustainable bison ranching to a discourse of settler nationalism that, he assumes, will cross-fertilize each other. Indeed, we are told from the outset, "the fate of the bison may well prove emblematic of the future of our nation" (1). It helps that the bison is easily romanticized according to well-trod tropes of America's nostalgia for itself. "Strength, endurance, adaptability, and cooperation in the face of danger make the bison a striking emblem of America" (2). Such masculine, no-nonsense, American cowboy boosterism appeals to the conservative midwestern establishment and right-leaning farmers, suggesting this vision of sustainability already conforms to their worldview and would not require any substantial change to their way of life (and appears as a patriarchal reversion in comparison to the examples of female empowerment in *Ecotopia*). "Bison are quintessentially American animals: stalwart, noble symbols of wildness, freedom, and self-sufficiency" (9). And if they weren't? What does this kind of nationalistic conservation rhetoric say about the sustainability of other animals? Yet Callenbach declares loudly only what other rewilding proponents tacitly condone: harnessing nationalism and animal nativism can be an effective way to repopulate animals, since these two agendas are likely to prove supportive of each other.

In proclaiming bison rewilding as a chance for American political and ecological restoration, Callenbach recognizes the bison's "absence is our loss, psychologically, spiritually, and morally" (16). He also adds, intriguingly, that the bison, which can weigh 2,000 pounds, "is the only large wild animal with whom there is any prospect of sustained coexistence on mass term" (16). Some have argued, even, that North American ecosystems miss the mammoth and camel and these, too, should be rewilded, with elephants as substitutes for mammoths.[15] The vision of large mammals restoring America has its own "go big" appeal, as if the return of giant animals could undo fossil-fuel dependency and more than a century of monocultural farming. "On the Plains we can transform current petroleum-based farming and ranching into an enduring, self-reliant system resting on the perennial resources of the region: sun, grass, and wind. On the Plains, a deep planetary challenge of long-term human survival waits to be met" (3).

Herds can graze under giant windmills, people and bison can live together, and new technology can converge with the ancient, prosperous ways of animal-human coexistence. Perhaps even what it means to be American will change. "If we decide that it is fitting for these noble beasts to share our future, and make room for them on the continent again, we will be a different people. It is worth entertaining the possibility that we will be a more humble, less driven, less exploitative people, with a livelier sense of connection to the wild in ourselves as well as in bison" (258). This rewilding endeavor appears to promise a moral as well as biological renewal, intertwining species and spirit. "Americans love happy endings, and the story of bison puts one within our grasp" (149).

This story of happy ends maintains a convenient routine for Americans and further ties the return of the bison within resurgent Plains-heartland nationalism that points to yet another enclosure within colonial history for the animal. Callenbach's American-booster version, while currying support from conservative ranchers that populate the Midwest, denies associating the return of the bison with efforts to dismantle the settler colonialism theft of lands and lifeways from Indigenous peoples. His manifesto is solicitous of Indigenous participation in bison ranching, but this argument does not include any mention of repatriating lands or reclaiming sovereignty for Indigenous communities and nations. Callenbach is certainly aware of the longer traumatic history of white settler treatment of bison, but he adroitly avoids any emphasis on this brutal legacy that has become part of the bison body. Instead, Callenbach calls for an ecological renewal as continuous with colonization and national forgetting.

In contrast to such settler nationalist rewilding claims to sustainability, the palpability of the return of bison has become an urgent project in the environmental and cultural restoration works by many Native American and Canadian Indigenous peoples in recent years. Ken Zontek remarks that "Native Americans have sought to preserve the bison as an extension of preserving themselves and their culture."[16] Winona LaDuke has written of the need to connect the history of curtailing where the bison roam and the many cases when the U.S. government reduced of the size of reservation land previously allotted to Indigenous groups.[17] Hence, as Nick Estes (Kul Wicasa) states, the Indigenous communities on the Plains have a "vital connection with the buffalo as sustaining continued Indigenous resistance."[18] An important initial step toward Native-led bison revival occurred with several Indigenous groups situated across Canada and the United States, signing "The Buffalo Treaty" in 2014, pledging to work together to connect the restoration of bison populations with the renewal of Indigenous cultural traditions.

Indigenous artworks that envision what it would be like to live with an abundance of bison in the present stress the importance of forging a renewed intimacy with the animal while remembering through trauma and reasserting self-determination apart from American hegemony. The bison depicted in literature and visual works by recent Indigenous artists—for example Thomas King's novel *Truth and Bright Water* (published in 1999), and Tasha Hubbard's short animated film *Buffalo Calling* (published in 2013)—draw attention to how the return of the animal links past and present in a landscape that is tense and politicized, but also musky, drenched in symbolism, and ripe for new collaborative human-animal and Indigenous futures. The artist Kent Monkman (Cree), in his acrylic painting *The Chase* (see figure 5-1), depicts bison in its post-extinction phase caught between a haunting and passionate existence. The painting shows a herd of bison stampeding through a nondescript,

FIGURE 5-1. Kent Monkman, *The Chase*, 2014, acrylic on canvas, 84" × 126", image courtesy of the artist.

rundown city street. The animals are pursued by an Indigenous hunter (a recurring character and alter ego in Monkman's work named Miss Chief), riding a rocket motorcycle, clad in sexy boots and carrying a Louis Vuitton-inspired arrow holder. Most of the running bison and bulls are depicted naturalistically, but some are cubist, and some are in the style of paleolithic cave art. Set in an unknown city, the painting brings together a striking combination of animal, technology, sex, art, primitivism, queerness, and Native cultural revival. The bison is brought back not in support of a nationalist-sustainability consensus but as a queer entity crashing through the streets, pursued by a gender-fluid, time-traveling, shape-shifting Indigenous hunter-artist. Monkman here depicts the bison as spilling into the decaying city and bringing together 30,000 years of bison art.

The return of the bison today shows the animal embedded in a world caught up in fetishizing the animal body, capable of seeing the bison amid other luxury goods and sensual consumer items. With bison running in paved streets and an Indigenous hunter shooting a bow while riding a motorcycle, everything seems a bit incongruent with place and time, yet the painting embraces living in a world full of anachronisms, inviting an ethos of rewilding, queering, and decolonization together. If the bison were to be brought back today, they would be entering a world where queerness and wildness mix, as Jack Halberstam argues— the wild as "a terrain of alternative formulations that resist the orderly impulses of modernity and as a merging of anticolonial, anticapitalist, and radical queer interests"[19]—rather than entering into the streamlined masculine bison-rancher bonds that Callenbach fantasized.

Monkman's bison painting connects an intense longing for the rewilding of the bison as inseparable from critiques of settler colonialism and foresees an expansion of experimental and aesthetic ways of being together with the animal in uncertain times. Bringing back the bison also means bringing back historical accountability for the traumas created by the decimation of the animal and the systematic dispossession and destruction of Indigenous peoples across North America. Dave Foreman repeatedly points to acknowledging "wounds" in the landscape as a necessary step to rewilding.[20] The point of bringing back the bison is not to institute a form of bison nationalism and capitalism but to change the definitions of sustainability to welcome new and various ways to desire and live with the animal. An Indigenous-led return of the bison certainly could be sustainable, as Callenbach imagines, and immersive, as Snyder implies, but the return of the animal requires fostering an understanding of the changing, ongoing relationships of Indigenous peoples to bison lives. After a deeply traumatic extinction event, the bison does not return simply as before; rather, the history of that extinction event remains in the very being of the animal. In Monkman's painting, the masculine, nationalist project of rewilding would become, instead, more diversely wilded in this queering of bison, sustainability, and Indigeneity together. The desire for the return of the bison, and a wilder biodiversity, takes place in a shaggy terrain of reconnected rural and urban geographies, promising more pungent political ecologies and muskier nature-cultures.

## The Tiger's Two Bodies

The global situation of tigers constitutes a near-extinction event, with current population numbers at about 97 percent lower than historical population levels. As with the state of bison, tigers have garnered outsized attention as charismatic large mammals, but this devotion has not resulted in consistent and unified efforts of conservation. Instead, tigers, like bison, have found themselves caught up in complicated networks of biocapitalism, nationalist agendas, and international conservation plans, in which practices of care and control overlap. Recent counts indicate that there are fewer than 4,000 wild tigers remaining (a slight increase from previous counts over the last decade). There are well over 10,000 tigers in captivity in North America—and probably a similar number in captivity across Asia—scattered in zoos and in private custody.

The reproduction and trade of domesticated tigers has little oversight in North America, and there exist tiger mills in China, Thailand, Laos, and Vietnam that breed the animals for tourism and to harvest their coveted pelts and body parts for their medicinal properties. Four of the eight subspecies of tiger have gone extinct in the past century. The remaining subspecies—Amur (Siberian) tigers, Indochinese tigers, Bengal tigers, and Sumatran tigers—each differ distinctly in size, coat markings, and genetic profiles. Most of the wild tigers are isolated in small groups on reserves or national parks. These tigers move about in forests and jungles exposed to poaching, logging, and game depletion. It will take a lot of monetary support, rigorous conservation efforts, and national and international cooperation just to stabilize and, perhaps, slightly grow the wild tiger population. The number of captive tigers is likely to increase with no oversight.

Tigers exist now as last animals and, due to the large number of them bred in captivity, also as uncounted, mass-produced domesticated animals. Tigers are classified as "endangered" by the IUCN, but this designation is specific only to wild tigers. Wild tigers in their own *in situ* habitats are nearing the precipice of extinction, while tigers *ex situ* continue to proliferate. Most domesticated tigers are heavily inbred and cross-bred among the different tiger subspecies. The genomes of these domesticated tigers are not consistent with the four remaining genomes of wild tigers and, thus, are deemed to have no conservation value (the Association of Zoos and Aquariums calls these animals "generic tigers"[21] and supports a breeding moratorium to halt their reproduction).

So far, there is no scientific evidence that a tiger bred and raised in captivity has the proper hunting skills needed to survive in the wild. The domesticated tigers in many zoos and private ownership, thus, have no apparent contribution to make to the survival of tigers in the wild. Although these tigers continue to reproduce and expand in numbers, their genome has no future outside the confines of captivity. However, the fate of wild tigers is closely connected to domesticated tigers, as both are highly prized in the exotic animal trade. The poaching of wild tigers is the primary cause of the plunging numbers that have brought the wild animal close to extinction. Both wild and domesticated tigers are understood by the broader public as exhibiting "tigerness," the natural and cultural associations of the tiger that contribute to its charisma and to the strong desire to possess and consume its body.

The disparity between domesticated tigers and wild tigers is so great they almost seem to be two distinct "genres" of tiger. This ontological schism raises a number of ethical, existential, and ecological questions. What, then, are the conservational ethics of last tigers compared to mass tigers? Which aspects of these two kinds of existence for the tiger should be definitive of "tigerness"? How can both wild and domesticated tigers be cared for as tigers when their "tigerness" is increasingly diverging? How have tigers become bifurcated into tigers that "count" and tigers that have no need of being counted? How can conservation of wild tigers remain distinct from domesticated tigers while acknowledging, at the same time, that wild and domesticated tiger cultures overlap?

The ontological bifurcations of these two genres of tiger raise several paradoxes of care. Conservation of *ex situ* tigers will not directly help the conservation of *in situ* tigers. Wild tigers exist at the limits of life in a situation that garners massive attention by local and international concerns, while domesticated tigers are too abundant to be counted. Care for wild tigers, however, is complicated by the difficulty of coordinating different agencies and communities whose interests in wild tigers do not always result in agreement on how best to protect the animal in the wild.

In India, where the largest numbers of wild tigers reside, Annu Jalais distinguishes between what she calls the "cosmopolitan tiger" and the "Sundarbans tiger" as two different frameworks applied to the *same* tiger in the Bay of Bengal region. The "cosmopolitan tiger" is the iconic animal for global wildlife agencies, an internationalized and often decontextualized animal, while the "Sundarbans tiger" is deeply enmeshed in local lifeways and disputes over land. Jalais remarks: "This global 'cosmopolitan' tiger, as opposed to the local 'Sundarbans tiger,' has become the rallying point for urbanites' concerns for wildlife protection."[22] Jalais indicates that the global, national, and local representations of tigers are each suffused with power dynamics that "perpetrate the coercive and unequal relationship between those who partake of the 'global' tiger view versus those who live with 'wild' tigers" (9). So much depends on what the term "wild" tiger, then, will designate. In India, tigers must be free and contained, nationalized yet sponsorable by international NGOs, regulated but postcolonial. In the context of bioconservation efforts that aim at preserving an animal's "wildness" above all, the philosopher Ronald Sandler points out that "preserving species will often require intensive management, thereby resulting in a diminishment in wilderness and naturalness (understood as independence from human design, control, and impacts)."[23] In yet another paradox of care, the more care for the wild animal, the less the animal is perceived as wild.

The associations of wild tigers with fierceness, primal predatoriness, and an instinct for survival in harsh climes are central features of tigerness that coexist uneasily with the reality of tiger conservation today, which depends on contributions from scientists, communities near tiger habitats, local trackers, and national and international financial support. John Vaillant's *The Tiger: A True Story of Vengeance and Survival*, provides an in-depth recounting of the pursuit of one particular Amur tiger that killed and ate several experienced hunters in the Primorsky region of the Russian Far East in the 1990s. Vaillant's title plays on the ambiguous referentiality of vengeance and survival, as both apply to tiger and human inhabitants of the remote, economically-limited area.

Vaillant tracks the history of the Amur tiger population decline along with the tumultuous Russian history in the twentieth century.[24] The Primorsky region, ceded to Russia by the Manchurian empire in 1860, was known primarily for its fur trade at the turn of the twentieth century. During Stalin's totalitarian regime, the region became a site for resettling displaced populations. The population of Amur tigers dwindled rapidly in the first decades of the twentieth century, as an estimated 100 animals were hunted each year in the 1920s. By the end of that decade, perhaps as few as twenty tigers remained. A few conservationists at the time raised the alarm for the species (Stalin cared little for the tiger), leading to the Soviet Union recognizing the tiger as a protected species in 1947.

Vaillant's narrative focuses on how, by the end of the twentieth century, some local hunters of the tiger continued to poach the animal while others sought to conserve it. The human inhabitants of the region, as well as the international community of NGOs invested in saving the tiger, each interpreted the tiger through their own objectives as a figure for survival in precarious times. Vaillant remarks that tigers "get our full attention. They strike a deep and resonant chord within us, and one reason is because, as disturbing as it may be, man-eating occurs within the acceptable parameters of the tiger's nature, which has informed *our* nature" (190–91).

Yet Vaillant's comment suggests a too easy conflation of endangered species survival with human survivalist rhetoric, a tendency that positions the Amur tiger within the framework of hunter/hunted that befits a romanticized frontier or post-apocalyptic mentality rather than community-focused conservation. The carnivorous tiger attracts so many paradoxes of conservational care worldwide because "there is no other creature that functions simultaneously as a poster child for the conservation movement and as shorthand for power, sex, and danger" (296–97). Vaillant does point out how such libidinal attractions to the tiger can be both helpful and harmful to the survival of tigers. In Vaillant's view, the only way forward for conservation seems to be fostering a complex collaboration among local and global interests in the animal while admitting that predatory vengeance/survivalist behavior among both humans and tigers will continue to play a role in the fate of the species.

Just 300 kilometers east of the Primorsky Krai region lies the city of Harbin in China, home to the Siberian Tiger Park, where hundreds of tigers of unknown genetic profile roam together in a penned area. Tigers are not normally social and do not spend time in packs. In the park, the tigers are on display to tourists who travel through the grounds by bus and can purchase meat to toss to the animals. There are, by estimate, over 5,000 tigers of mixed genealogies kept across fourteen registered tiger farms and parks in China.[25] Although China has signed on to the CITES (the Convention on International Trade in Endangered Species of Wild Fauna and Flora) ban of wild tiger international trade, the agreement allows for the loophole of domestic, captive-bred tigers to be considered legal commodities within the confines of national trade.[26] Tiger body parts are highly prized items in traditional Chinese medicine. "Though none of the big cats' parts have been shown to have any medical value (placebo effect excluded)," Rachel Love Nuwer remarks, "pretty much everything has been assigned a use. Whiskers quell toothaches, meat cures malaria, fat stops vomiting, blood strengthens willpower, noses sooth children's epilepsy, teeth purge sores from a man's penis, eyeballs and bile prevent convulsions, and penises banish impotence and promote longevity. Of all the tiger's parts, though, its bones are the most sought after."[27]

The legal market for tiger parts in China is just a small part of the larger market that extends across South and East Asia for exotic animals considered to be traditional medicines or delicacies. According to statistics gathered by CITES, between 1998 and 2007, over 35 million exotic animals regulated under the agreement were legally traded in the region, including 16 million seahorses, 17.4 million reptiles, 400,000 mammals, and 300,000 butterflies.[28] The majority of these animals were caught in the wild. While this trade often is stamped as legal, the illegal harvesting of wild animals is frequently mixed in, using the legal trade as a cover. The value of legal trade of wildlife has been calculated at $323 billion in 2009 by TRAFFIC (a wildlife trade monitoring NGO), while the illegal wildlife trade is estimated to be a $7 to $23 billion annual business by the United Nations Environmental Programme. Although tigers cannot be traded internationally, the mixing of wild and domesticated tigers in lucrative operations feeds what Vanda Felbab-Brown calls an "extinction market." The tiger farms participate in larger and overlapping licit and illicit marketplaces in which the commoditization of exotic life leads to the situation in which the animal, in the phrasing of Jacqueline Schneider, is "sold into extinction."[29]

It is important to understand and critique treatments of the tiger in marketplaces situated in China and elsewhere in the region while also rejecting racist and derogatory designations of the Asian "Other" often predicated on discourses that animalize and demean such communities (similar pejorative language is attached to local communities across the globe that struggle with efforts at preservation of endangered species while addressing daily needs for communal sustenance and well-being). Felbab-Brown, in her study of the international exotic animal trade, documents how wildlife markets for endangered animals, including for tigers and tiger parts, have grown in locations across Asia, Africa, North America, and Europe through confusing legal and illegal sources of income. Dismantling the trade in endangered animals will require adaptable programs that emphasize conservation using enforcement methods such as trade bans in connection with support for local anti-poverty initiatives, community empowerment (such as ownership in local ecotourism economies), and social justice efforts. In many cases, an alternative and synthetic material can be substituted for the coveted animal product (such as artificial keratin for rhino horn), to help stem trade in the animal while recognizing the reality of market demand. She remarks it would also be more feasible to reduce the consumption of tiger parts by supporting larger financial commitments to affordable community medicine on a national and international scale. Felbab-Brown stresses that conservation efforts must be continually vigilant and flexible, since "What works today in one place may not work tomorrow in the same place, and it may already not work today in another place . . . There is no policy silver bullet to stop the bullets of poachers."[30]

The overlap of extinction and marketization for tigers takes a different form in the United States, where tigers are in many states legal to breed in captivity. The animals are sold to circuses and private zoos, and are coveted by private purchasers as animals that confer special stature and powerful freedoms. Camilla Calamandrei's documentary film *The Tiger Next Door* provides a window into the realities of domestic-born tigers by following the tiger breeder Dennis Hill in his legal and financial battles to continue to keep two dozen tigers captive on his private home grounds.[31] Over a period of several decades, Hill had bred tigers and kept them enclosed in small chain-link fenced cages, about ten by twenty feet in size, on his rural property in Flat Rock, Indiana. Although Hill repeatedly called the tigers his "children," he sold the cats in a lucrative business to a variety of buyers, including magicians, private zoos, and circuses. Hill mentions that the big cats could fetch up to $15,000, and adds that white tigers, created by inbreeding, might sell for as high as $150,000. Hill's permit to keep captive tigers had come under review in 2005 by the Indiana Department of Natural Resources, after an inspection revealed the big cats in their small cages in what was evidently miserable conditions. Ultimately, Hill got rid of all but three tigers; however, his right to own and breed tigers in the first place was never questioned at the judicial level.

Owners in North America of these exotic animals repeatedly drive home the point that the freedom to possess these animals must not be infringed. The biopolitical fate of these North American tigers is closely tied to the current spectrum of political rhetoric associated with libertarian claims for maximum individual human freedoms with minimal government oversight. Hill bemoans that "once again the government has taken something from me" and insists "it's deeper than me and tigers— it's about freedom." Going as far as to declare "I'm living as a tiger myself," Hill superimposes his own thwarted libertarian claims onto the caged lives of his tigers (a similar defense of tiger captive ownership as synonymous with extravagant displays of American freedom appears in the cult hit film *The Tiger King*, released in 2020).

Ironically, the tight, painful confinement that tigers must endure in the care of these private owners is the very means by which the owner performatively asserts his freedoms. Caged tigers that must submit to the whims of their owners carry enormous economic and symbolic capital as objects of exotic wealth, personal power, carnivorism, virility, and anti-government demonstration. The lives of tigers often are caught up in conspicuous displays of masculinity and assertions of individual prerogative regardless of public risks or the animal's own interests.[32] The endangered status of tigers amplifies their exotic allure as highly sought after "lively capital," in the phrasing of Rosemary-Claire Collard, but also can further endanger these animals' lives. Collard's work documents how, for exotic captive animals, "the processes of enclosure, individuation, and control that form the exotic pet as capital are driven by the pet's value as a lively, encounterable being, yet these very processes diminish and often even end life for exotic pets."[33]

Tiger breeders can make significant profits from the animal, whether alive or dead. Declarations of care or fondness for the animal by these breeders can quickly become subordinate to individual whims of boom-and-bust libertarian capitalism. Hill's case is an example of what Sarah Jaquette Ray describes as an instance of "packaging . . . hidden attachment to white supremacy in animal love."[34] The welfare of the tigers is at the caprice of Hill's ideological and economic dynamics that are encoded within longer racialized histories of liberty as the right to own and dominate nature.

Even as captive tigers and wild tigers lead increasingly bifurcated existences, their lives remain intertwined, as the work of bioconservation cannot be neatly separated from biopolitical and biocapitalist situations at individual and national levels. The reality of life at the edge of extinction has served to intensify the overlap between the care and commoditization of tigers. The lives of wild and domestic tigers are intertwined with nationalistic identifications, exemplifying of what Neel Ahuja calls the "government of species" according to which some species find their lives regulated according to "biocapital investments in national, racial, class, and sex factors."[35] Tigers are iconic figures for extinction today as much as they are figures for Russian toughness and predatoriness; Chinese independence and economic prowess; American personal liberties, which include the freedom to own and display conspicuously exotic animals; and India's efforts to chart its own postcolonial path on issues concerning conflicts between land conservation and development. The

much sought after "tigerness" of the tiger is now inextricable with different individual, governmental, and nongovernmental modes of attention
and care that require ongoing critical scrutiny.

## Coral Cultures in the Anthropocene

Corals are rapidly joining the unfortunate ranks of charismatic megafauna facing extinction in Anthropocene times. Because many cannot
survive after just one degree Celsius average of ocean warming, Corals
are one of the most vulnerable species on the planet. Coral reefs, which
can be visible from outer space, befit the status of megafauna although
they are made up of millions of microfauna.

The moniker "charismatic megafauna" has been applied usually only
to large animals deemed "flagship species," mostly mammals, which have
been taken up by conservation groups as icons for the urgency of saving
endangered species.[36] The phrase, however, has accrued a moderately
negative connotation recently, as many scholars now use the term "charismatic megafauna" to recommend against the over-investment of conservation research and attention on just a few favored species.[37] These
attractive megafauna are seen as crowding out care and funding for the
millions of other species on Earth. The focus on megafauna seems not to
be ecologically rational, since it does not account for species interdepen-
dence, complex food chains, and distributed ecosystem-building processes. But while these critiques are warranted, charisma still can remain
ecologically and culturally relevant by fostering emotional connection
alongside empirical environmental science to maintain both passionate
engagement and critical watchfulness for endangered animals. If humans
cannot commit enough conservation to maintain charismatic animals,
what chance does the rest of life have? A critical reclaiming of the discourse
of charismatic species can provide analysis of the problematic effects of
narrow conservation favoritism while still considering how fantasies and
emotional attachments can be effective in soliciting plural forms of environmental care.

Coral already has a long association across several centuries as a
source of charismatic enchantment and wonder, as Marion Endt-Jones
discusses in *Coral: Something Rich and Strange*.[38] Until recently, most of
the cultural associations around coral emphasized the exotic biological
and sensual qualities of coral, which is composed of animal, mineral,
and vegetable (algae) elements. Coral has fostered fantasies of floating
wonders and dreams of alternative underwater lifeworlds, but also figures in histories of shipwrecks and the precariousness of humanity at

sea. Darwin studied coral as a robust maker of atolls that bedeviled the burgeoning shipping industry, while Herman Melville, in *Omoo: Adventures in the South Seas*, wrote about coral stringing the Pacific Islands as glittering "marine gardens" of "coral plants of every hue and shape imaginable."[39]

Oceanographer Roger Revelle, who participated as a scientist in nuclear tests on coral atolls, stated in a 1954 paper: "Of all earth's phenomena, coral reefs seem best calculated to excite a sense of wonder. And of all the forms of coral reefs, the atolls have appeared to men of science to be the richest in mystery and the most strange."[40] Though Revelle also studied the effects of climate change on oceans, he thought coral maintained itself in patterns of disturbance and recovery, and that reefs were resilient and sturdy as a whole, since they are capable of building and sustaining atolls thousands of feet tall in the ocean. This paradigm of resilient coral was replaced by the results of research done in the 1980s and 1990s that documented increasing waves of mass coral death, called bleaching events. The cultural and ecological associations of coral shifted toward considering reefs as particularly precarious, sensitive, and acutely exposed to three of the most rapidly changing aspects of the Earth's climate: rising temperatures, sea level rise, and ocean acidification. Coral's multiple vulnerabilities and perilous conditions have become now enmeshed with its previous characterizations of enchantment and strangeness.

In recent decades, as evidence of mass bleaching events increasingly has been registered by marine scientists, coral is fast becoming a barometer for biodiversity loss and rising rates of species extinction. "Reefs are disappearing before our eyes," marine biologist Rebecca Albright states. "In the past 30 years we have lost approximately 50 percent of corals globally, and researchers estimate that only about 10 percent will survive past 2050."[41] Just as coral is often considered to be the "rainforest" of the sea, as it harbors a major portion of the ocean's biota, the death of coral reefs entails staggering losses of life across many ecosystems and food chains.

To understand coral today, then, requires attention to situations of enchantment and extinction at the same time. Coral has long figured as the site of intermixing of life and death in metaphorical associations of skeletal bone and luscious garden. In the Anthropocene, coral continues to be the bearer of such associations but now with planetary-scale implications for all life. Stefan Helmreich points to the urgency to make coral legible in different ways: "Coral is something to be read—for climate change, for potentially patentable genes, for representativeness."[42] Coral's charismatic life and charismatic death bring together nearly all the key material and conceptual issues of the Anthropocene: species extinction, multispecies interconnections, global warming, ocean acidification from carbon pollution, agricultural and industrial pollution, global fisheries and destructive industrial fishing practices, global tourism, Indigenous island and coastal communities, and contemporary art's continued fascination with coral's precarious otherness. One way, then, to think about the entanglements of the Anthropocene, extinction, and capital, involves thinking with and as coral.

Recent work in the environmental humanities has emphasized the difference between thinking *with* as compared to thinking *about* another nonhuman life. To "think with" involves approaching nonhuman lives with a sense that curiosity and care are not just epistemologically wise ways of knowing these lives but, also, are the means to foster recuperative practices as collaborative among multispecies. Donna Haraway borrows the phrasing "becoming-with" to propose how "becoming-with is how partners are . . . rendered capable"[43] and to "learn how to conjugate worlds with partial connections and not universals and particulars" (13). The preposition "with" features in the work of several environmental humanities scholars (for example, the volume *Thinking with Water*, edited by Cecilia Chen, Janine MacLeod, and Astrida Neimanis[44]) who advocate that modes of thought need to expand beyond the empirical or data-driven sciences that tend to reaffirm human control and ownership over nonhuman lives and landscapes. They encourage ways of analyzing and relating to the nonhuman that include an openness to different cultural modes of knowing that involve local knowledges, decolonial commitments, and new ways of aesthetically engaging with nonhuman lives.

By examining works of coral art alongside a reflection on coral's charismatic status in the Anthropocene—a practice of "thinking with"—I draw out some of the ecological and cultural concerns implicated in how coral is becoming "the ocean's canary," in the phrasing of acclaimed marine biologist J. E. N. Veron.[45] If coral is treated under the metaphor of the barometer or as analogous to the miner's canary, what kind of ecological status or charisma is implied? Justin Prystash comments that, because coral "straddles the organic and inorganic and constitutes a series of connections between individual polyps, limestone, barnacles, fish, and other marine life, coral lends itself to metaphorical proliferation."[46] Contemporary artistic representations of coral strain to reconfigure the plenitude of metaphorical associations of coral as caught between fragility and resilience.[47] To think about coral and coral art together requires connecting the multiplicity of coral metaphors with careful attention to multispecies lived realities in the present. As I will detail, coral's metaphors are rapidly becoming extinction's metaphors.

Existing at the juncture of rock and organism, geology and biology, coral reefs are constructed by individual saclike polyps that gather together into colonies (though some polyps can live individually). While not a plant, corals are often symbiotic with photosynthetic algae. Incorporating zooxanthellae or tiny colored algae gives coral its bright colors. The algae provide the coral nutrition from photosynthesis that is put into the service of building a calcium carbonate structure. Not all coral polyps build reefs (of the approximately 2,500 known coral species, 1,000 are hard corals), but those that do pull the raw materials of calcium carbonate from the ocean and secrete them as an exoskeleton. Reef-building corals grow primarily in warm waters (though some reef-building species exist in cool deep-ocean waters as well) within a narrowly defined range of acidity and at shallow depths, since the algae need access to sunlight for photosynthesis.

Assemblers of aquatic complexity, corals are multispecies beings that are archipelagos for other multispecies populations. Corals often are called "underwater rainforests" in that they provide harbor for a tremendous amount of biodiversity (the Great Barrier Reef, with 400 different types of coral, 1,500 species of fish, and 4,000 mollusks, has more aqueous biodiversity than all of Europe[48]). It is estimated that one-quarter to one-half of all aqueous life depends in some degree on coral reefs.

As polyps, symbionts, and colonies, corals are multiple beings interdependent with multiple temporalities and multispecies relationships. Corals build communities on the death of previous iterations of itself; a coral colony can survive for several hundred years on reefs that are 10,000 years old, but that reef could have been formed by offspring from a previous reef. Reef-making coral date back 500 million years, while other shell producing aquatic animals date to over 700 million years. As with ice cores, scientists have drilled through coral reefs to help compose the climatological history of the planet. The advantage of using coral is that it provides a much longer history of sea level that can be correlated with temperature changes and greenhouse gas cycles. These core coral samples have provided important data for charting the planet's previous mass extinctions, which further associates the history of coral with the understanding of planetary extinction processes. Veron calls corals "messages from deep time,"[49] while at the same time, he points to how rapidly coral is changing today, remarking that it may be within the timespan of only a century that much of coral life will be eradicated.

The phenomenon of mass coral bleaching has been increasing in intensity and frequency in recent decades. In a bleaching event, the algae leave the coral host, which turns bright white in color. The coral is not dead yet, but in a highly stressed condition. While some corals are able to recover, many succumb to disease or lack of nutrients. The local extirpation and global possibility of extinction for corals would initiate a cascading biodiversity loss among other species. The mass bleaching events across the planet, thus, indicate a new condition for coral and its symbionts, transforming these "gardens" from sites of enchanted fantasy and biodiverse plenitude into sites of dystopia and dread.

The primary charismatic characterization of coral as "something rich and strange" is morphing into a phrase for the Anthropocene. Coral biology has historically informed two types of metaphor: one drawing on its similarity to the human skeleton and the other on its radical alien otherness. Shakespeare's *The Tempest* captures these opposite associations in what is commonly known as "Ariel's Song." The spirit Ariel misleads the shipwreck survivor Ferdinand into thinking that his father did not survive the boat's sinking and that his body has transformed into coral at the bottom of the sea:

> Full fathom five thy father lies,
> Of his bones are coral made.
> Those are pearls that were his eyes,
> Nothing of him that doth fade
> But doth suffer a sea-change
> Into something rich and strange. (I.ii 397–402)[50]

The conversion of human bone into coral is a sea-change that is possible because of the perceived similarities of the two substances. The lifeless human body transforms into living coral, following along conduits of the imagination's anthropomorphic tendencies and tales of species metamorphosis.

Yet at the same time, everything remaining of the human body that lies on the sea floor undergoes a radical change into "something rich and strange." The becoming-coral of the body transforms into something precious and valuable (pearls) as well as uncanny and haunting. Shakespeare's lines show how coral's rich and strange ontology beckons anthropomorphism, yet also, coral is a kind of life that commands wonder precisely because reefs appear to be an entirely independent underwater civilization rarely seen. Shakespeare's poem also hints at the interconnected ontologies of human life and ocean life, such that sea-change signifies an interactivity between the two domains. The acceptance of such a sea-change is evident even in recent medical research developments that use coral as material for bone grafts in human patients.[51] The sea-change scene in Ariel's song is embedded in a work that borrows from accounts of European colonization of Caribbean islands. Thus "sea-change" carries with it a history of violence and the middle passage. The phrase has come to refer to the devastation of ecologies after centuries of settler colonialism.[52] Now "sea-change" also signifies for climate change and species precarity.[53] The lives of animals in the sea are undergoing a sea-change in a time of global heating—for that matter, even the sea is undergoing a sea-change.

In Darwin's early cogitations on coral in his *Notebook "B"* (dated 1837–38), he expresses the notion that coral communities might provide the most apt model for thinking about the interconnections of all life. "The tree of life should perhaps be called the coral of life, base of branches dead; so that passages cannot be seen–this again offers contradiction to constant succession of germs in progress."[54] Darwin rejected his own idea for this metaphor to visually model the origin of species, but as Justin Prystash notes, these early writings prompted reflections on how coral played a prominent role in Darwin's thinking that lead to his questioning of "dominant conceptions of subjectivity, gender, and time."[55]

Darwin viewed corals as powerful builders and "myriads of architects"[56] that could construct works of art vaster and more sublime than humans could. "We feel surprise when travelers tell us of the vast dimensions of the Pyramids and other great ruins, but how utterly insignificant are the greatest of these, when compared to these mountains of stone accumulated by the agency of various minute and tender animals!" (465). It is not clear if Darwin is winking at the reader, since the Egyptian pyramids, as well as the Mayan pyramids, are made of limestone sediments compacted in large part from ancient coral reefs. Darwin's admiration at the "accumulated... agency" of reef-building corals contributed to a trend in the nineteenth century in viewing the collective activity of coral as a kind of communal political labor. In Jules Verne's *Twenty Thousand Leagues under the Sea*, the narrator remarks on how corals possess "their own existence while at the same time participating in communal life. They, thus, live a sort of natural socialism."[57] Counter to doctrines of individualism and notions of endless strife among living beings, coral came to be seen as embracing communal entanglements. Yet this sense of coral as having its own independent political life is now being superseded by the global biopolitical entanglements that coral finds itself caught up in.

These references to coral as a special kind of oceanic community now need to be contrasted with discussions of coral signaling the broader collapse of oceanic communities. In more recent attention to coral, the collective life and death of coral has featured as a measure of the precarity of oceanic conditions in the Anthropocene. The metaphor of coral as "ocean's canaries" applies to both the endangerment of coral communities and the human communities that depend on the biodiversity supported by coral reefs. In addition to the "ecosystem services" of reefs as nurturing fish and crustaceans that provide primary food sources for hundreds of millions of humans, corals serve as both material and symbolic resource for many coastal and Indigenous island communities.

Indigenous peoples in the Pacific, the Indian Ocean, and the Caribbean have formed extensive and intimate connections to coral. Native Hawai'ian communities hold the belief that the birth of an island, including its coral, is the manifestation of "the most ancestral being"[58] where the sea, sky, and land come together to supply the necessary ingredients of life. The *Kumulipo*, a Hawai'ian chanted poem composed in the early eighteenth century, opens with the creation of life from coral-covered islands. A coauthored study by scientists and native Hawai'ian ethnographers details the Hawai'ian view that "the skeletal composition (calcium carbonate) of corals and humans is the same material, further supporting the belief that we are direct descendants of corals" (107). The researchers add, "Corals are a foundation of life in Hawai'i. Throughout the Native Hawaiian understandings of natural processes, intimate, reciprocal relationships between man and coral have been nurtured in both spiritual and practical realms" (114).

The Tonga scholar and writer Epeli Hau'ofa has called for a solidarity among Oceania peoples rooted in the stewardship of the Pacific in concert with other coastal Indigenous communities across the globe. He states, "No people on earth are more suited to be guardians of the world's largest ocean than those for whom it has been home for generations."[59] The life and death of coral overlaps with Indigenous struggles for decolonization and oceanic solidarity in the face of rising seas. To think with coral then is to recognize its intertwined conditions with coastal and island Indigenous life, the *longue durée* of Native ways, and contemporary movements for Indigenous protections of land, food sovereignty, and political self-sovereignty.

The multispecies and multicultural communities associated with coral now face accelerated precariousness as evidence of mass bleaching events becomes more evident. Today, it is estimated that 75 percent of all coral reefs are threatened in their near-term survival. The combination of enchantment and extinction that describes the status of coral communities also applies to understanding recent works of art that aim to provide visual and material exemplification of coral's changing fate. Certainly, living coral is extremely photogenic, but recent mass bleaching events also have shown how the death of coral is also visually striking. How, then, should one look at coral death? How does one understand the visual evidence of such death and what kinds of reactions and responses can be drawn from such exposure? The documentary Chasing Coral,[60] directed by Jeff Orlowski, grapples with the problem of how to visualize and react to coral's spectacular, even charismatic, death. The film follows a small team of coral activists in their attempts to provide photographic evidence of the rapid onset of bleaching at several reefs around the world. The documentary provides viewers with a brisk education in the biology of corals, the crisis of bleaching reefs in warming and acidifying seas, and the difficulties of constructing reliable underwater cameras.

Most of the first half of the film centers on the work of diver Richard Vevers in his visits to several dive sites along with coral scientists and his determination to provide time-lapse photography of coral bleaching on a large scale. The second half of the film shifts the focus to one of the underwater camera crew operators, Zack Rago, who reveals his childhood devotion to coral and becomes the film's emotional focal point as he struggles to witness and photograph a massive bleaching event in the Great Barrier Reef during the 2015–16 Australian summer.

The devastation that year to the Great Barrier Reef received considerable press, including one example of journalism that took the form of an obituary and eulogized: "The Great Barrier Reef of Australia passed away in 2016 after a long illness. It was 25 million years old."[61] Though the whole reef has not succumbed, making this elegiac announcement awkward and controversial, this kind of pronouncement, along with documentaries like Chasing Coral that show what the death of a reef is like, highlight the need to find a provocative form and mode of address to reckon with the wider impact of coral's precarious condition.

Australian poet Judith Wright writes: "If the Barrier Reef could think it would fear us," and adds "we have its fate in our hands."[62] Much of the documentary's narrative is preoccupied with how to make the film and present coral's story, hence the "chasing" title. The various successes and failures of equipment and storytelling provide an example of the modernist formalist technique of "laying bare" the device to let viewers know we are not witnessing via the unmediated eye of nature but with highly mediated and limited camera technology. The vast majority of contemporary nature documentaries seek to naturalize the artifice of the camera lens so the viewer feels as if the animal is intimately right there before the gaze. In the Planet Earth series (2006, produced by Alastair

Fothergill), for example, the camera and filmmakers are never to be seen, and while no humans enter the visual space, the voice-over from David Attenborough is performed as if he were present in the moment watching the animals directly.[63] By contrast, *Chasing Coral* treats the physical and emotional struggle with the cameras as integral to the authenticity and reality of the story of trying to provide a reliable and intimate witness of the life and death of coral. The strategy of discussing the film as one is making it also is intended to direct the viewer to see *Chasing Coral* as aware of its straining to articulate a pathway that includes scientific documentation, environmental activism, and entertaining visual drama. The film, ultimately, composes these perspectives into a work that suggests the emotional seesaws and self-aware mediations of these ways of engaging with coral will provide a model for how to form a global coral culture movement.

The film crew eventually manages to provide single stills and time-lapse photography of the brilliant colors and liveliness of coral as well as the dismalness of what mass coral death looks like. In reckoning with the visual sea-change of "something rich and strange" into dead coral, Vevers provides a voice-over narration that serves as another moment of "laying bare" when he states, "you see a picture of a beautiful white reef—is that a good thing or a bad thing?" With the explanation that we are watching luminous white coral in its dying throes, the film then assumes typical expectations of nature documentaries as it teaches the audience how to correlate seeing and feeling. In trying to guide viewers in how to react to this "incredibly beautiful phase of death," Vevers slips directly into the voice of coral as he tells viewers that "it feels as if coral is saying 'look at me, please notice.'" In this moment, coral's death, an event of the loss of animacy, is felt to be personified. As this comment arrives near the end of the film, both filmmakers and audience are thrust into a bitterly dissonant experience of simultaneous thrill and devastation at being able to successfully see the Great Barrier Reef's mass destruction. The accomplishment of obtaining images of coral bleaching becomes both the emotional high and low of the film, a simultaneous success and failure.

The film's technological achievement of underwater still shots taken day after day over several months provides a before-and-after-shot experience that solicits further reflection. In the final minutes of the film, viewers see several before/after images of corals in different locations, and have the ability to slide between the time of life and death. In one glance, one sees some of coral's multiple temporalities and possible futures. The slimy waters left after coral's demise add yet another visualization of gothic decay to the larger catalogue of destruction in the Anthropocene.

The natural-cultural meanings of this kind of double vision is generally legible as a dichotomy of good/bad, life/death, utopia/dystopia. But this film's ability to provide visual evidence of the collapse of biodiversity prompts viewers to expand the meaning of before/after photography by contextualizing these images as part of the larger visual repertoire of climate change linked to extinction events in the present. Thus, the before/after photographs, such as the view of a staghorn coral colony (*Acripora muricata*) in American Samoa succumbing and becoming covered in algae (see figure 5-2) also are meant to provide evidence of the before and after of warming and acidifying seas. Here, we see biological time come into direct conflict with Anthropocene time, with the speed of climate change overwhelming the slow capacity of coral to adapt. The before/after also provides a visual analogue to the viewer's changed relationship to a new knowledge and responsibility—now that the audience knows the precarious state of coral, one cannot insist that the globe will see the dire effects of climate change only much later in the future.

FIGURE 5-2. Richard Vevers, Staghorn coral in American Samoa, before and after, photograph, 2014–2015, OceanImageBank.org.

Observing coral gardens turn into algal slime calls for a reimagining of the oceans as rapidly trending toward "gelatinous futures," in the phrasing of Stacy Alaimo.[64] "The Anthropocene seas will be paradoxical, anachronistic zones of terribly compressed temporality where, it is feared, the future will move backwards, into a time when the oceans were devoid of whales, dolphins, fish, coral reefs, and a multitude of other species, but jellyfish (and algae) proliferated . . . [T]here is little enthusiasm for a future in which the oceans become like the ancient acidic seas, characterized as 'slime-rock systems'" (158). Understanding coral in the Anthropocene necessitates collapsing present and ancient temporalities, revealing a time in which complex life forms retreat as earlier cellular life seems to advance.

The before/after shots of coral collapse, however, also are meant to mobilize a new kind of global communal coral activism. In the final minutes of the film, coral biologists and Indigenous island coral watchers from around the world offer brief testimony of the dire conditions of coral in geographically specific locations across the planet. Yet, up to this point, the film did not provide any interviews or commentary from global coral communities, including Caribbean, Pacific Islander, and equatorial coastal peoples, for whom coral is intimately intertwined with ancestral and contemporary lifeways. The film provides only a brief nod to how the loss of coral also entails the loss of coral cultures and traditional lives bonded to coral's condition.

Even as the film employs the device of double vision to make legible coral's transition from life to death, the challenge remains for viewers to disrupt the linear, oppositional structure of before/after by striving to imagine and realize alternative paths and futures for coral. Coral's next sea-change must occur even upon this current moment of mass coral death. Bärbel Bischof has highlighted how coral scientists face multiple difficulties in assessing and preserving Marine Protected Areas, including the high cost of research, disagreements in academic methodology, and the recurring problem of sidelining long-standing local and Indigenous stewards of coastal biomes.[65] Irus Braverman, in her book *Coral Whisperers*, details the flourishing research into assisted migration of resilient corals and artificially constructed reefs. Braverman pursues the enmeshed emotions of hope and despair of coral scientists as they measure disappearing coral life and attempt to regrow reefs using techniques that include artificial coral reef transplantation and breeding hardier corals. As Braverman suggests, "immersing ourselves in the multiplicity of coral life allows us to step back and recognize the many assumptions . . . that underlie our understandings and our regulation of life and death—both in our environment and, eventually, in ourselves."[66]

The mystique of coral, which is similar to that of tigers and bison, attracts new conservation and awareness efforts to think-with and act-with animals on the brink of the existential threshold of extinction. These animals have become central causes for developing new forms of cross-species care that require continually renewed critical attention. Either the total loss of bison, tigers, and coral or their widespread repopulation would reverberate massively throughout the meanings of extinction. The return of these animals would signal that something had gone tremendously right in efforts to stem global heating and habitat destruction. It also would mean shifting away from paradigms of animal capital that thrive on rarity and captive exoticism that assume an unquestionable liberty and access to the buying and selling of animal lives. Bringing back these animals from the brink of extinction will involve new dependencies and shared precarities, along with new forms of social, scientific, and aesthetic attention toward thinking and being with nonhuman lives.

# 6 /  What Is De-Extinction?

In a 2013 TED talk by Stewart Brand, now viewed over 2 million times, Brand presented a sweeping vision of how de-extinction could change life on Earth. "What if you could find out that, using the DNA in museum specimens, fossils maybe up to 200,000 years old could be used to bring species back. What would you do? Where would you start?"[1]

Brand debriefed the audience on the basics of de-extinction science and how the implantation of embryos of either frozen DNA or reconstructed DNA of extinct animals into living hosts can bring the lost animal back to life. Declaring that "conservation biologists are realizing that bad news bums people out," Brand proclaimed de-extinction as a bold step toward optimism and concrete results in bioconservation. Brand extolled his own Revive and Restore foundation dedicated to bringing the passenger pigeon out of extinction, and added that he expected similar good news in the near future for extinct Carolina parakeets, great auks, heath hens, ivory-billed woodpeckers, Eskimo curlews, Caribbean monk seals, and woolly mammoths. Turning to the audience, Brand cheered them on: "What do people think about it? You know, do you want extinct species back? Do you want extinct species back?" As the audience clapped, Brand suddenly morphed into a Peter Pan figure, announcing that "Tinker Bell is going to come fluttering down. It is a Tinker Bell moment."

Let's recall that, in J. M. Barrie's play *Peter Pan*, Tinkerbell drinks a poison meant for Peter to save his life. As Tinkerbell's life force fades, Peter breaks the fourth wall of the stage and calls to the audience: "She says—she says she thinks she could get well again if children believed in fairies! Do you believe in fairies? Say quick that you believe! If you believe, clap your hands!"[2] The applause revives Tinkerbell in a moment of theatrical triumph in which fairies and fictional characters share the same existential possibilities as children and the play's attendees.

What can this scene tell us about the possibility of de-extinction? Brand's appeal to the audience for applause is meant to stimulate both the audience's imagination of what it would be like to encounter a de-extinct animal and also as an indication of approval for developing the biotechnological means to bring such animals back. Applause leaves no room for questions, critiques, or debates about the nuances, problems, and potentialities of this decision. In the play, the audience's applause passes back through the fourth wall of the theater and includes the playgoers in the play. Audience participation encourages everyone to re-enter the theatrical space of fantasy, belief, and childhood wishes come true.

Brand has in mind an analogous effect in which the audience of the TED talk applauds the bringing together of fantasy and biology. Clapping is somehow to be directly connected to the new biotechnologies—applause made flesh. But wishing for extinct animals to come back, even fantasizing about it, is not the same desire as actively entering the laboratory to recreate animals through biotechnology. Brand's encouragement of magical intervention toward making animals savable has offstage consequences the audience has not been encouraged to consider.

As we see in this chapter, the desire to bring back extinct animals has implications that go way beyond resurrecting a handful of recently extinct species. De-extinction has the potential to be a "this changes everything" technology. This is not just about bringing back recently lost species; such salvific biotechnologies would redefine extinction and the existential conditions of finitude for all species. Yet, as this chapter discusses, this resurrection of animal life, which promises to right past wrongs inflicted on extinct animals, exemplifies how care for the existential aspects of animal life is not exempt from the biopolitical manipulation of bodies toward ends that do not necessarily lead to more ethical relationships with animals or better ecological practices overall.

De-extinction technologies implore us to ask, again, the question "What is extinction?" across all the facets of life and death. The implications of de-extinction science also reverberate with the imaginative scope of science fiction populated with extrapolations on how developments in genetic science might fundamentally change the categories of humanity and animality. Genetic modification, cloning, and the storing of genomes in gene banks already are reconfiguring definitions of species and extinction. The scientists and conservationists who are promoting the use of genetic science to revive lost species consistently claim that the return of these species would be ecologically beneficial. They also claim that humans have a moral obligation to revive species whose extinctions were caused directly by human predation or indirectly by habitat destruction. However, discussions of de-extinction that are limited to dilemmas in conservation do not fully reckon with the fundamental existential changes heralded by the advancement of such technologies.

De-extinction, if successful, would remake the meaning of extinction. The science of de-extinction must be understood in the broader context as a "power over" definitions of life and death. The ramifications of developing new powers over life and death resonate across the existential conditions of both humans and animals. Such power requires examining how ecological concerns about conservation can shift quickly into the most dire questions concerning how much control humans should have over life itself and how the crisis of extinction can be used to legitimate such powers. As we will see, reviving life, increasing life longevity, and enhancing life are conjoined techno-utopian aims that must be understood together as controversial new forms of power and value.

This chapter also provides a reading of Octavia Butler's *Xenogenesis* trilogy as a science fiction allegory that extrapolates on how de-extinction technologies might apply to humans as well as animals. Butler's collection of three novels, *Dawn, Adulthood Rites,* and *Imago,* (also published together with the title *Lilith's Brood*), appeared in the late 1980s in the context of the rise of genomic technologies that began to point to ways of doing speciation and making generations in the lab. Butler's trilogy imagines possible de-extinction futures that intertwine utopian and dystopian scenarios and do not resolve into any easy position for or against de-extinction. These novels connect the power to de-extinct life with issues of genetic determinism and what it means to maintain or lose control of the reproductive power of one's genome. Butler's trilogy aligns science fiction and the future of science in a narrative concerning how to think and act self-reflexively about extinction in the era of the Anthropocene, in which all life has become by definition precarious life.

De-extinction sciences have advanced quickly in recent years. The most widely celebrated example is the case of the Pyrenean ibex (also called the bucardo), a wild goat native to the rocky Pyrenean mountains. The Pyrenean ibex population was decimated in the last century by human predation. The last Pyrenean ibex died in 2000. In 2003, scientists in Spain, using genetic material harvested from the last few bucardos, cloned the DNA to create a number of copies of the nucleic material. From one of the cloned nuclei, a viable embryo was produced, which developed into a newborn bucardo brought to term in the womb of a domestic goat. The newborn, however, survived only seven minutes, succumbing to a respiratory problem from under-developed lungs. This case is credited as the first animal to become de-extinct, although many scientists dispute proclaiming this instance a success since the bucardo did not survive long. The Pyrenean ibex currently remains extinct, but this first effort to return a species into existence using contemporary genetic science provided tangible evidence that extinction may not be forever.

There are three main genetic engineering technologies that can potentially be used in de-extinction: back-breeding, cloning existing DNA of an extinct animal, and reconstructing DNA of an extinct animal by modifying similar existing DNA of a living animal or recreating the extinct DNA using nucleic building biotechnologies.

Back-breeding, previously discussed in chapter 4, works by selectively breeding existing animals that exhibit some of the physical and behavioral characteristics of a closely related species, with the intention of bringing those characteristics to the fore. The extinct animal is not wholly revived but some of the genotype and phenotype of the missing animal can be reestablished, with the result that an approximation of the lost species can appear to be present.

Cloning, used by the Spanish scientists, proceeds by somatic cell transfer. An intact sperm or egg cell must be taken from the still-living animal and stored cryogenically. The nucleus of that cell can be implanted into a host receptor cell and can, then, be brought to term in the uterus of a closely related species. The newly born animal will have the same genome as its parent, essentially being a physically identical animal. This technology works only for animals from whom embryonic cells have been collected while the animal was alive. Since collection of these cells for select endangered animals began only in the late 1970s, this technology would not apply to any animal that went extinct before that time period.

A third technology involves reconstructing a full or partial genome of an extinct animal in a lab and implanting the nucleic material into a cell that is brought to term by an animal with a similar gestation cycle as the lost animal. This form of de-extinction technology is necessary when there is no viable cryogenically preserved embryonic DNA, such as the passenger pigeon and the mammoth. Each of these technologies requires further analysis into what Charis Thompson calls the "ontological choreography"[3] of assisted reproductive technologies (ART), which extend across and remake the human–animal divide and lead to what it means to live a de-extinct life.

## Cryogenic Conservation

Gene bank collections of endangered or extinct animals are often called "frozen zoos." A frozen zoo is a collection of sperm, egg, and skin cells taken from a species and preserved in tanks of liquid nitrogen at -196 degrees Celsius. The first facility of this type was developed in 1975 by the San Diego Zoological Society led by Kurt Benirschke.[4] This research center, which copyrighted the name "Frozen Zoo," now has samples from more than 1,000 vertebrate taxa. Frozen zoos have garnered significant research support and funding momentum in the wake of increased attention to the evidence of a sixth mass extinction, including a major facility recently established in England called the Frozen Ark. The metaphor of the salvational ark or zoo gives these projects a millenarian halo, what Donna Haraway describes as a "sacred-secular" narrative of "both physical and epistemological rescue"[5] that belies the difficulty these research centers face in achieving reproductive success for animals rapidly dwindling in numbers.

Oliver Ryder, director of the Frozen Zoo in San Diego, urges conservationists to "look beyond crisis-oriented recovery actions"[6] and support cryopreservation of endangered species. Ryder contrasts the present biodiversity crisis with the deep-time archival potential of frozen zoos, stating that this technology could "enable preserving forms of life as banked cells for centuries, millennia, or even geological epochs" (261). Ryder does note that some may be concerned with how gene banking could become a form of bioprospecting. He concludes, however: "In an era of declining biodiversity and expanding capabilities in genome biology and genetic editing biotechnology, a new relationship with nature stands to emerge" (266). The phrase "Frozen Zoo" is copyrighted by the San Diego Zoo, which has prevented other gene banks from using the moniker. Here, then, for free use are some other names: Cryozoo, Cryogenic World, Cryonic Park, Frozen Bank, Ice of Life, Cold Company.

Frozen zoos store the germ line of an endangered species in a suspended state and, as the metaphor of the "bank" indicates, the species will be "redeemed" at a later date. But gene banks and frozen zoos already have effects on what Ryder calls "a new relationship with nature," including redefinitions of conservation and extinction in the present. Gene banks contribute to shifting conservation emphasis from the organism-plus-environment unit to the microbiological-archival unit. This technology changes the primary theater of ecological management from the outdoor habitat to the space of the lab and its cold containers. The endangered animal undergoes a dramatic ontological change in cryopreservation as indicated by the paradoxical term "suspended animation." The embryo and gamete cells are as much technological lab materials as they are biological reproductive materials, as much inert objects as they are live subjects.

Joanna Radin and Emma Kowal call cryopreserved life "latent life, the liminal and vague state between life and death."[7] Cryopreserved germ cells are not living and are not dead. This liminal state raises questions concerning to whom these cells actually belong. Stored in patented banks and proprietary labs and zoos, the cells owe their origin as much to the work of genetic scientists as they do to the reproductive lives of animals. Nikolas Rose discusses how in modernity the category of "life itself" has become multiplied and redefined along increasingly technological and economic lines. "Life itself has been made amenable to these new economic relations, as vitality is decomposed into a series of distinct and discrete objects—that can be isolated, delimited, stored, accumulated, mobilized, and exchanged, accorded a discrete value, traded across time, space, species, contexts, enterprises—in the service of many distinct objectives."[8]

The divvying up of life itself into "discrete objects" reinforces the idea of animals as collections of disparate parts that can be frozen, stored, and manipulated as needed. Cryogenic preservation and cloning are inextricable from ways of producing new kinds of "biovalue."[9] For example, germ cells can be sent to different labs and across borders in ways live animals cannot. ART, as all technologies, are prone to obsolescence and in constant need of technological upgrades and ongoing institutional funding.

Research into ART for endangered species dovetails with well-funded research into ART for the industrial animal food market and human infertility treatments. The increasing technologization of sexual reproduction for humans can provide for more freedoms and control for persons to choose how they reproduce (thus supporting the feminist aim of women having reproductive control over their own bodies). But these same technologies also can further biopolitical and economic agendas involved in commoditizing and optimizing preferential forms of life. As Michelle Murphy has observed, "Sex's changeability expanded further, beyond humans, to intensify in the animal and plant kingdoms as agribusiness mutated seeds into patentable commodities, and livestock was bred with artificial insemination and embryo transfer. This rapidly emerging technical ability to alter human and nonhuman reproduction, stretching from molecular to transnational economic scales, was accompanied by new problems and promises for the politicization of life—not just should, but how could reproduction be transformed?"[10]

The existence of cryopreserved germ cells of last animals implies that these species are not extinct, but these cells are not living animals, either. In response to the rise of de-extinction technologies, Douglas Ian Campbell and Patrick Michael Whittle claim that a new category should be added to the IUCN Red List classification system: terminally extinct.[11] This category would indicate that there would be no scientific means to revive the species (for example, there is no way to reconstruct dinosaur genomes because DNA degrades immediately after death and has a half-life of 521 years). In the case of cryogenic conservation, if an animal has its genome stored in a bank, it would not be "terminally extinct."

However, this redefinition of extinction categories reinforces the case that compiling and storing information, be it encoded in a cell or in a computer, can serve as a placeholder for biodiversity. Matthew Chrulew discusses how this attitude toward bioinformatics as a proxy for conservation fundamentally can change definitions of extinction: "As information can be discovered and stored in sites other than bodies, or in bodies in states other than living, a species might thus be located beyond even the last survivor of its kind. This relocation of life onto DNA that can be stored and manipulated in computational or cryogenic archives made the previously impenetrable void of extinction a knowable and potentially redeemable domain."[12]

As Chrulew and others have noted, cryopreserved cells and lab-based de-extinction work neglects how each species exists in an interconnected habitat in a specific time and place that includes ongoing relationships with other species, including humans. These relationships cannot be frozen and rethawed when convenient. Cryopreservation science also bypasses the varieties of local knowledge and Indigenous practices of care for cohabiting with animals and ecosystems. Advocates for cryopreservation, however, argue that archives function as time capsules designed to leave traces of the present for the future, planning for time scales longer than a human generation, which is essential in preparing for long-term ecological problems. Frozen zoos anticipate a potentially catastrophic futurity and a technologically advanced and biodiversity-caring future civilization that can put such an archive to use.

## Reconstructing Extinct Genomes

The de-extinction technology receiving the most attention today involves attempts to recreate the partial or full genome of an extinct animal. For this technology to work, scientists must first locate enough existing DNA from an extinct animal to create a comprehensive map of its genome. The mapped genome of the extinct animal can, then, be compared to the mapped genome of the closest living ancestor. Using specialized tools for cutting into DNA strands and replacing genes, such as CRISPR/Cas9, the sections of DNA of the living ancestor that differ from the extinct animal can be isolated, snipped out, and replaced by the DNA coding of the extinct animal. The genetically engineered genome then would be placed into an embryonic host cell and brought to term by the closest living ancestor. The success of this process requires tremendous feats of bioengineering to go right at each stage. For comparison, in the cloning of the bucardo, scientists had made 285 embryos and only one of those came to term resulting in a live birth.

If all these bioengineering steps prove successful and result in a live birth of the extinct animal, there are still several more steps to go toward the survival and ultimate de-extinction of the animal. The newborn must be raised in such a way that the infant gains the knowledge and habits learned from its parents that are required for survival. Since no parents of the species exist, scientists will have to find ways to simulate that education. The success of raising one newborn is not enough. Other newborns with a wider range of genomic profiles must be added to prevent inbreeding and to contribute to adaptive behavior at the level of the group.

Finally, the de-extinct species must be able to inhabit a supportive ecosystem. The ecosystem that originally formed the animal's habitat will have changed, if not wholly disappeared. The threats to extinction remain, and the de-extinct species will face new interactions with other species. The de-extinct animal will need to forge new relationships with ecosystems, other species, and humans to rejoin shared biotic and cultural communities. The difficulty of each of these steps is enormous, and each stage could prompt complications or failures that could be due to technological, biological, ecological, or cultural reasons.

What would success for a de-extinct animal look like? Beth Shapiro, in *How to Clone a Mammoth*, states adamantly that the de-extinct animal would not be identical to the extinct ancestor. There currently exists no 100 percent complete genome of any non-cryopreserved extinct animal. Scientists may be able to approximate some of this genome for extinct animals deemed desirable to revive, but any genetically engineered animal constructed from these remade genomes will differ to some degree from its extinct ancestors.

The IUCN, in a 2016 report on "guiding principles" of conservation for de-extinction, calls such remade species "proxies," as distinct from cloned "facsimiles."[13] In the case of the mammoth, there will be no 100-percent revival of the extinct animal. Even discoveries of frozen mammoth bodies with skin and hair provide only partially recoverable DNA. The only pathway to de-extinct the mammoth is to modify the Asian elephant genome to express traits characteristic of the mammoth, though this modified genome may be only a degree of 1 percent closer to the mammoth genome.

Since genetically and behaviorally exact revivals are not possible, Shapiro observes: "This provides an opportunity to redefine de-extinction, shifting away from a species centric view."[14] Some would be content with calling the hybrid elephant-mammoth a "mammoth 2.0" (it would be more accurate to identify the animal as a genetically engineered Asian elephant), but this is still to focus on the re-making of the original species. Shapiro argues, instead, that the importance of de-extinction is in restoring lost ecologies and ecological relationships. "In my mind, it is this *ecological resurrection*, not *species resurrection*, that is the real value of de-extinction. We should think of de-extinction not in terms of *which life form* we will bring back, but *what ecological interactions* we would like to see restored" (131).

Ben Novak, who leads work at Revive and Restore on the passenger pigeon, views de-extinction also as foremost an effort of "ecological replacement" by "purposefully adapting a living organism to serve the ecological function of the extinct species."[15] He adds, "The goal of de-extinction is to restore vital ecological functions that sustain dynamic processes producing resilient ecosystems and increasing biodiversity and bioabundance" (5). Shapiro and Novak insist that de-extinction should align with re-wilding efforts, although it bears noting that the re-wilding projects that presently exist are able to do their work without depending on expensive, speculative technologies, exclusive patents, and currently nonexistent animals. As discussed in chapter 5, as the norms of wildness are changing, there is no clear consensus on what re-wilding entails. For some, it includes market-driven solutions that can make repopulation dependent upon tying the animals to economic and nationalistic agendas. For others, re-wilding is an ethical and biodiverse project that must foreground the need to grow animal populations in the context of environmental justice movements based on community consent, decolonization, and democratizing public space. Most of the efforts of de-extinction rely on a techno-centric fix that evokes bioethical re-wilding rationales loosely. This technology tends to reinforce decision-making on animal lives by isolated, well-funded elite individuals and institutions whose work will have enormous effects on the public sphere but may not end up methodologically incorporating public consultation or consent, among other ecojustice practices.

The work of biodiversity conservation today does not need to be cast as technophobic or purist; it already relies on a variety of assisted reproductive technologies and extensive human management of animal lives and habitats. Novak points out as much by saying: "There is no species alive today that has not adapted in some way to human activities of the past and thus become changed from its pre-human contact state" (8). However, the argument for de-extinction technologies as supplying the key missing piece—the extinct animals themselves—in conservation ecology is highly questionable. Most species that are candidates for de-extinction have long been absent from their habitats, which have now changed significantly. If the desire is to change these ecosystems back to previous states of biodiversity abundance, there are other methods and existing species that can satisfactorily restore such ecosystem conditions.

Many conservation ethicists have voiced the concern that de-extinction technology is a distraction from conservation field work and can siphon away much needed time and money from more practical and less invasive conservation methods. There also is a massive difference between assisting existing animals and creating a new kind of animal, for example, an elephant-mammoth, a mammoth proxy, or mammoth 2.0. Since these new hybrid animals are not identical to the extinct animal, there is no way to atone for the wrong done to animals that have been needlessly and purposefully rendered extinct by direct or indirect human actions. De-extinction is not justifiable as an obligation to render justice for previous harms since the revived animals are proxies at best, nor is there an obligation to bring new, hybrid life forms into the world. "Extinction, in other words, *is* indeed forever," Curt Meine remarks. "De-extinction, it follows, is a literal impossibility. The products of de-extinction can never be more than artificial proxies."[16]

In Ronald Sandler's assessment of the ethics of de-extinction, the best case for de-extinction is neither the duty to right past wrongs nor a determination to restore long-lost ecosystems. The best case is rather that successful de-extinction would create new cultural, technological, and economic value, while providing a counternarrative to environmental decline and loss. "[D]e-extinction would be a tremendous scientific and technological achievement," Sandler remarks. "Accomplishing it would require advances in genetics and synthetic biology, among other fields. It would likely spin off further research programs, technologies, and applications. . . . Moreover, many people would find it wonderous and awesome to see a living ivory-billed woodpecker, thylacine, or mammoth, even if only in a zoo or wildlife park."[17] The desire for an achievement on the scale of the technologically sublime, rather than the animal's own welfare, is granted priority in this view. The "wow factor" Sandler recognizes as a justification hints that the real payoff of de-extinction likely will not be foremost in the service of bioconservation but in the technosphere.

Perhaps the most substantial effects of de-extinction would be in the development of new existential technologies aimed at human life extension and transhumanist enhancements. The field of de-extinction science includes a number of celebrity scientists who see themselves as entrepreneurs in the mold of tech moguls. One of the most pivotal scientists involved is Harvard University geneticist George Church, who has made it clear that he sees de-extinction as one tool in the larger toolbox of technologies to pursue longer lives and genetic enhancements for animals and humans alike. The potential success of using genetic technologies in bringing back extinct species is a stepping-stone toward applying these technologies to human life toward transhuman (humanity augmented by science and technology) objectives.

The same tools that can modify an animal genome can modify a human genome, leading to improved physical powers and health prospects. Church encourages genetic science to go much further than this. He extols the possibility of "civilizing, taming, and domesticating the basic processes of life"[18] and "maximizing evolution" (89). Church aligns this work with the aims of the transhumanist movement to achieve maximum life longevity and the pursuit of continual upgrades in physical and mental capabilities. This pursuit of enhanced and potentially endless life will necessitate eventually leaving Earth, given all the eventual climatological and astronomical perils facing the planet. Church declares that "we should develop equipment for rapidly detecting and deflecting such events and/or moving some of our civilization out of the way, and off the planet" (244). Writing of what he calls the "panspermia era" (225), Church foresees the genomic descendants of humans as cosmic wanderers: "The genome should become not just the genome of one lonely being or one planet. It should become the genome of the Universe" (244).

In Church's view, genetic technologies should pursue both species reconstruction and enhancement, blurring the line between de-extinction and transhumanism. Church makes no distinction between developing bioethically convincing medical therapies using genetic modification (for example, eradicating malaria from mosquitos or extinguishing the Ebola virus) and the biotechnological modification of the body to thwart any mortal risk. The use of the term *regenesis* in the title of his book integrates conservation and life extension in an Earthly and cosmic triumph.

Many Silicon Valley titans are avid sponsors of research into life longevity as tantamount to extinction prevention. Larry Page and Sergey Brin, co-founders of Google, aspire to "cure death" (the stated aim of their start-up company Calico). In Nikolas Rose's analysis, effacing the lines between biotechnologies of life enhancement and the treatment of mortal illness is a hallmark of contemporary biopower that "is enthusiastically engaged with the biological re-engineering of vitality."[19] As a result, "The old lines between treatment, correction, and enhancement can no longer be sustained" (17). It still remains important, however, to make the case that the pursuit of maximized control over evolutionary and reproductive processes is not the same as biodiversity conservation.

The desire to live an enhanced transhuman life that persists into deep-time futures is not the same as the concern for sustaining the deep-time of Earthly ecologies. The claims for more life, longer life, and revived life each raise distinct concerns about the biopoliticization and capitalization of vitality. While transhumanism deserves much more scrutiny than this chapter can provide,[20] the motive to maximize life continues to draw from the technologies and discourses of de-extinction, yet shifts the purposes for such work further away from the concerns of how to find a way to share the planet ecologically with diverse life forms.

## De-Extinction as an Existential Technology

An analysis of the ethics of de-extinction as a contribution to biodiversity is not broad-based enough. A more thorough comprehension must properly reckon with the existential implications of this technology. While de-extinction is declared to be in the service of conservation ecology, these technologies initiate a wholesale set of changes to the definition of ecology. By claiming to act foremost in the name of biodiversity, scientists and entrepreneurs supporting such technology deflect from the larger political and ontological questions that this technology unleashes. It is crucial to understand how de-extinction would change what ecology means by changing what life and death means. As discussed in the introduction, while too much extinction undoes the ecological condition, no extinction also undoes the ecological condition. There is no such thing as an ecology that does not include the possibility of being otherwise or being finite. De-extinction technologies would fundamentally redefine existential precarity and finitude and, thus, usher in something radically different than what currently constitutes species existence.

Deborah Bird Rose's conceptualization of what she calls "ecological existentialism" helps us further understand how de-extinction technologies will redefine ecology. In her book *Wild Dog Dreaming: Love and Extinction*, she claims that "Ecological existentialism thus proposes a kinship of becoming: no telos, no *deus ex machina* to rescue us, no clockwork to keep us ticking along; and on the other hand, the rich plenitude, with all its joys and hazards, of our entanglement in the place, time, and multispecies complexities of life on Earth."[21] Ecological existentialism recognizes the dynamic quality of relationships that are constantly changing, such that "nature" is not a fixed reference (and not opposed to technology) but an ongoing exchange committed toward the support of entangled life. This existentialism is not an anthropocentric humanism but a devotion to explore and experiment within the "joys and hazards" of the existential condition as intertwined with biodiverse Earth systems. Such ecological existentialism is immediately accessible as a multispecies commonwealth and not predicated on expensive industries and salvational technologies. Rose's existential ecology welcomes not the overcoming of extinction but a process committed to constantly reaffirming a shared ecological commons.

De-extinction at first seems like a minor emendation to how ecosystems function, since the technology currently applies to only a few cases of lost animals. But once the core features of irreversible ecological change are under our control—such that we can make and remake species and ecosystems because we no longer view the death of an animal as a limit or a problem—then our definitions and foundations of ecology will have to be drastically revised. De-extinction is said to contribute to ecological conservation, but by fundamentally changing the parameters of ecology, there seems to be no way to effectively evaluate the claims of ecological benefits provided by this technology.

The intertwinement of speciation and extinction at the core of Darwin's thesis of evolution would be rendered irrelevant if de-extinction became a predominant technology (the primary principle of selection would be more Lamarckian since "acquired" or manufactured genetic characteristics would be passed on). These technologies bypass interdependent relationships and what Rose calls the "ancestral power"[22] of generations that comprise ecosystems. De-extinction is, thus, not re-ecologizing but de-ecologizing. De-extinction substitutes place-based ecological contingencies and relationships with programming problems and top-down decisions over biopolitical values. For some, this is a feature, not a reason to halt such technologies. Perhaps in the Anthropocene it is too much to ask for ecosystems to function primarily by existential-ecological means, rather than, first and foremost, through technology. It is crucial, however, to develop further comprehension and critical reflection on what existential technologies are and how they change definitions of life and death, along with ecosystems and extinction. We have just begun to understand extinction and confront its consequences; now, we are seeing the terms of extinction undergoing radical change due to the emergence of new technological powers over the lives of species.

What kind of power, then, is it to have power over extinction or de-extinction? What is power at the level of the species that can make decisions on the very presence or absence of the species form? The power to de-extinct is connected to the long history of what Michel Foucault calls biopolitics, political power "applied at the level of life itself."[23] While the terms of Foucault's biopolitics have become very familiar and, perhaps, too obvious in analyses and criticisms of how the biological condition has been incorporated into the political condition, they remain cogent and helpful in understanding the implications of life transformation technologies. As a counterpoint to the "wow" factor of de-extinction efforts, it is important to comprehend these new existential technologies and transhumanist agendas in the context of the conceptual and political history of confronting biological limits as sites of crisis and control.

Foucault names biopower as a "set of mechanisms through which the basic biological features of the human species became the object of a political strategy, of a general strategy of power."[24] Biopolitics is the "matter of taking control of life and the biological processes of man-as-species and of ensuring that they are not disciplined, but regularized."[25] Foucault tracks the rise of modern biopolitics, starting in the eighteenth century, focusing on the intersecting discourses and institutions invested in "regularizing" life by codifying sexuality, reproduction, mortality rates, familial structures, and population health. These modernizing knowledges of the body reciprocally interacted with new biological understandings of life and death, racial difference, gender norms, health protocols, and species categories. Cumulatively, these biopolitical macro- and micro-powers used to classify and regulate bodies and populations span the entire cycle of life, hence Foucault's summative statement that biopolitical decisions wield "the power to foster life or disallow it to the point of death."[26]

It is not the case that biopolitical practices wholly subsume the realities of life and death into the totalizing convergence of biology and power; these existential categories also are "lived experiences" and serve as horizons of possibility and finitude that can be socially and creatively transformed toward more careful ends. However, biopolitical theory stands in critique of traditional existentialism that validated ahistorical and universally applicable conditions of human existence and "being-toward-death" that presumed an unquestioned and unmarked notion of "life" and its norms as guiding precepts. Foucault's historicizing and critiquing of life and species concepts disrupted the existential faith that "life will find a way." While techniques of the discipline and regulation of "life itself" are applied across all subjects of modern political spectrums, the intensification of biopolitical doctrines since the eighteenth century have contributed to the establishment of norms of the human as primarily white, able-bodied, economically productive, and properly reproductive. Bodies marked as different from norms in terms of race, ability, gender identification, and citizenship status experience matters of life and death not just as existential conditions provoking a generic anxiety of finitude but as contested everyday realities.

In the contemporary politics of nature, the fact that essentially all animals are legally and conceptually treated as different from and subordinate to humans functions in continuity with biopolitical definitions of the preferential status accorded to these normative conceptions of life, race, and species. Foucault did not examine how the human/nonhuman animal divide has become a predominate norm (what is called "speciesism" or anthropocentrism), and Foucault's own limited Eurocentric and anthropocentric purview of biopolitics has been well critiqued by others.[27]

In the modern era of biopolitics, some human and animal bodies are prioritized and optimized, while others are marginalized, exhausted, subjugated, left to die, or intentionally killed. The biopolitical management of human life is fundamentally intertwined with a similar management of animal life across the planet, ranging from agriculture to the use of animals in commodity production, pet care, and ecosystem services models of investment in wildlife sustainability. Animal life is subject to a wide variety of administrations and interests that intertwine forms of care and concern as well as control and dominance. The differential values assigned to animals is historically intersectional with measuring systems that rate the animalization of humans along lines of racial difference. As Claire Jean Kim stresses, "Race has been articulated in part as a metric of animality, as a classification system that orders human bodies according to how animal they are—and how human they are not—with all of the entailments that follow."[28]

De-extinction raises new concerns about the range of powers over the metrics of species as well as the limit conditions of life and death for human and animal life. The historical iterations of biopolitics that Foucault identified largely assumed human populations as already identifiable and categorizable and did not seek to transform the very biological basis of life and death. The modernization of biopolitics has required the maintenance of life and death as governable and coherent designations. It is one thing to politically manage the health of the people—divvying up social rewards along the lines of the normal and the abnormal, favored and un-favored, with bodies cast according to metrics that follow from intersect-ing racial and animal orders. It is another thing to seek control over what constitutes the very power of living and the threshold of dying.

To make the distinction most relevant here, biopolitics historically has been concerned with decisions that prevent or condone extinction that have followed primarily from metrics of the racial and animal hierarchies of dominant interests but not with resurrecting animals and not with changing the ontological condition of humans and nonhuman animals such that extinction is no longer an existential limit (even though, as dis-cussed in chapter 4, the Third Reich sought to redefine extinction and practice de-extinction, they did so not to abolish death as such but to control it for their own ends). Whereas biopolitics historically has been concerned with who gets to share the Earth, as Arendt phrased it, de-extinction points to the power to make alterations to the existential structures of humans and animals that would lead to redefinitions of what counts as extinc-tion and ecology. Because these redefinitions of the entire existential structure from birth to life to death facilitated by de-extinction tech-nologies are not evidently motivated by democratizing social and eco-logical justice but are, rather, made with the view of immense profits and personal benefits even as they announce a new era of animal care, they require further critical analysis as especially transformative forms of biopolitical power.

There has been a recent push in discussions of biopolitics in Anthro-pocene times to place the older Foucauldian model under further scru-tiny to account for new theorizations of animal personhood, racial justice, and the risks and potentials of new existential technologies (including genetics and artificial intelligence), with planetary implications. Eliza-beth Povinelli questions if the biopolitical methods and characterizations listed by Foucault can fully account for the new powers over what consti-tutes matters of life and death in what she labels "settler late liberalism." As "the previously stable ordering divisions of Life and Nonlife shake," there emerge "new figures, tactics, and discourses of power"[29] that dis-place the older biopolitical paradigms. Povinelli then reasons that "a new drama" is being played out in the biopolitics conjoining settler colonial-ism and neoliberalism, a drama that includes power over "Nonlife" and determinations on "the extinction of humans, biological life, and, as it is often put, the planet itself—which takes us to a time before the life and death of individuals and species, a time of the *geos*" (8–9). Povinelli coins this kind of dominant paradigm as geontopower, which connects power over what gets to be called living or extinct to power over the geological condition of the entire planet. While modern biopolitics addresses humans and nonhuman animals at an individual or species level, geontopower is the planetary-scale power to decide what constitutes life and nonlife in the first place—for example, to decide if rivers, trees, corporations, artificial intelligences, or the entire Earth itself have the status of persons or living beings.

This power over the ontological condition of the entire Earth—the power to set or change the material and metaphysical rules for the existential structures of life and nonlife—is exemplified in the casting of anthropocentric sovereign authority over the totality of the planet. These arguments for engaging with a new kind of biopolitics of the Earth coincide with recent theorizations in critical race studies and Indigenous studies—including writing by Achille Mbembe, Alexander Weheliye, Tiffany Lethabo King, Zakiyyah Iman Jackson, Glen Sean Coulthard, and Jodi A. Byrd—that examine how white settler power is perpetuated by organizing ontological definitions and thresholds in a way that casts Black, Indigenous, and racially marked bodies as incessantly subject to erasure and finitude.[30] Oppressive power over the distribution of the terms of humanity and animality along the lines of racial and species metrics continues to coincide with the power to determine what counts as a "lived experience" or nullity of existence, and what the very terms of life and death, generation, and extinction mean.

These insights into this new level of species power and geontopower provide crucial guidance in how to understand the reach and transformation of de-extinction technologies today at a planetary level. The biotechnological ability to control the parameters of species finitude must be understood as powers implicating all life on Earth at an existential level that are fundamentally intertwined with issues of justice and colonialism. These technologies make extinction—formerly a certainty for all species and the basis for a shared precarious condition—a more mobile and manipulatable concept that is, at the same time, more determined and dependent on technological upgrades and entrepreneurial human interests that have no apparent bases in ecojustice principles.

Throughout this book, extinction has been defined as an existential-ecological condition in which the natural and cultural meanings of extinction are enmeshed in experiences and concepts of limits and lastness that are historically specific to different times and places. De-extinction technologies will, ultimately, lead to a severing of these existential and ecological interrelations. Such technologies point to a future that is de-ecological and de-existential, because they would replace the precarious and finite interrelated conditions of existential ecology with a view that these conditions are programmable, controllable, and improvable at will.

Reviving a handful of species is just the beginning for this technology, which is an early piece of the transhumanist puzzle toward new kinds of radical life preservation, longevity, and enhancement. These technologies, ultimately, aim to make the risky and existential processes of birth, life, care, and death increasingly pliable, marketable, and upgradeable. They view resurrected life, enhanced life, and transhuman life as the highest biovalues. But this kind of power over existentiality itself evidently is not the pathway toward a more just and shared planet, as such technologies and those who sponsor them favor personal and entrepreneurial benefits that are not particularly driven by responding to the history of social justice and anti-oppressive movements in support of developing networks of care and solidarity across human and animal differences. There is no evident passage from de-extinction labs and transhumanist entrepreneurs to community-based ecojustice.

At this juncture, one could say in defense of de-extinction: It makes sense to be skeptical of genetic power that extends into existential power. But let's be honest and admit that if an extinct species could talk, surely it would want to be revived. What if we sanction de-extinction for recently demised species but stop the technology at that point? Cloning is commonly practiced in agriculture, and seed banks have wide participation among international partners who gratefully send their plant species for preservation. Genes are just codes that offer varying affordances and limitations for an organism. Modifying existential coordinates by storing, reviving, or changing the genetic code is not "playing God" but, rather, tinkering with a program that is already designed for plasticity and adaptability. Modifying the genes of other life forms is nothing new—we need to remember that humans have been using artificial selection processes on plants and animals for thousands of years.

This argument continues: De-extinction technologies such as gene banking, cloning, and genetic modification so that species can persist in a tumultuous time are methods of "last resort" that we should cultivate "just in case." Just letting species disappear, even though we have the technology to bring them back, is a moral failing, as well. The writers of the IUCN *Guiding Principles* document already recognize that the environmentalist commitment of preventing extinctions is not the same as the commitment to de-extinction: "Humans have a moral obligation not to render species extinct, but it is unclear if this extends to a moral obligation to resurrect them."[31] However, we do have a moral duty not to ignore ways to stem the rising rates of extinction. This does not lead to the obligation to condone all de-extinction technologies, but technological assistance in stemming extinction should be welcomed.

What these kinds of arguments still miss are the biopolitical and colonial histories embedded in the very accolades over fostering life and determinations on death that continue to marginalize and exhaust some lives over others. Historically, the obsessive pursuit of more and more "life itself" over the past few centuries has coincided with the devaluing and effacing of lives marked as racially, sexually, aesthetically, and economically inferior. As Matthew Taylor recounts, "Historically and conceptually, then, there is no 'life' without racism, no vitality without hierarchy."[32] De-extinction technologies likely would have no ameliorative effect on such oppressions in the name of "life" that intersect across human-animal conditions. They fail to incorporate environmental justice and social justice practices at the initial methodological level. The ethics of species revival also cannot be separated from the wholesale transformation of the very ethical and ecological models that are used to assess the decisions made using these models.

In thinking through the implications of de-extinction science, it matters that we recognize the technological revivification of species as an option and not an obligation. The philosopher Hans Jonas, an early proponent of bioethics who describes his work as "an 'existential' interpretation of biological facts,"[33] argues in *The Imperative of Responsibility*, that humans are not responsible for perfecting themselves or their world. Jonas reckons with the planetary scale of world-making and world-destroying technologies (similar to the diagnosis of geontopower made recently by Povinelli) by stating that humans are obligated to care for the future effects of their actions and for the future of the plenitude of life on the planet. But humans do not have an ethical imperative to achieve perfection or dissolve suffering. Adapting Jonas's argument, animals should not go extinct—we have an obligation do everything to prevent this from happening to currently living species—but there is no corresponding obligation that all animals must be saved from mortality as such or granted maximum or eternal life so long as technology allows. Ethics is the granting of care to the precarious, not granting perfection to the imperfect. As Jonas states, "Only for the changeable and the perishable can one be responsible."[34]

A critical reflection on de-extinction as having "this changes everything" effects on ecology and existentiality does not mean one is committed to an anti-technological organicist or naturalist worldview. It is possible to recognize ecological conditions as enmeshed in historical, conceptual, and technological processes that define planetary existence today and, yet, still refuse technological interventions into the ecosphere that will fundamentally change the existential parameters of life. Amy Lynn Fletcher describes these de-extinction technologies as producing "life-on-demand": "the twenty-first-century environmental imaginary, in which life is made to order through mastery of technology and seemingly fixed borders such as the line between extant and extinct become much more porous and open to human intervention, is firmly rooted in aspirations to exert control over nature that characterize the modern biotechnological project."[35]

This view of life ties de-extinction on demand with remaking ecosystems on demand. The phrase "on demand" denotes that human willfulness and interest is prioritized over the demands of animals. One possible scenario of de-extinction as "life-on-demand" would result in a situation where animals go in and out of extinction due to technological availability and popularity trends. For example, passenger pigeons may pass through extinction more than once, as the pigeon's population moves in yo-yo fashion, exiting extinction when there is enough funding and attention and falling into extinction yet again when these resources wane or interests are exhausted. De-extinction technology aimed at reversing finitude has its own finitudes.

Life in the Anthropocene is marked by the reality that it takes hard work to maintain a shared planet—hard work that should be collective and equitable, and that should not exhaust ourselves, the animals, or the ecosystems of the planet. However, we should not need to commit to new ontological technologies of resurrection "on demand" in order to care for that life. De-extinction has the potential to break the existential links between life, care, and finitude shared among humans and nonhuman species. Care matters because the conditions of life are not permanent or guaranteed. Care includes the concern for one's own possibilities and finitudes, but, also, care entails immersion in the world and the relationships that comprise our shared condition. Care also includes anticipating the loss of something that matters, and committing it to memorialization once what matters is lost irretrievably.

While de-extinction might point the way toward undoing the loss of recently lost species, it also undoes the fundamental conditions of care. Assisting the flourishing of human and animal lives does not require technological control over their existential structure. Never letting go is not the same thing as sustainability. Although promotors of de-extinction claim to be acting in the name of biodiversity and animal welfare, this technology puts the animal being further under a form of human control that is not necessarily aligned with supporting biodiverse life in the context of maintaining the work of care and the pursuit of environmental justice on a shared planet. The leaders of de-extinction technologies emphasize imagining what it would be like to meet a mammoth, but they have not expressed interest in issues of working to dismantle oppressive biopolitics that associate racialized peoples with animal behavior, nor do they foster any particular emancipatory or decolonial aims such as animal liberation and restituting land to Indigenous peoples who have generational, place-specific connections to animals.

S. D. Chrostowska observes that, "In its current forms, desire for immortality has been written by the history of domination."[36] The refusal of the cult of de-extinction technology—the physical and epistemological mastery of life and death, the capitalization of finitude into new biovalues, and the transformation of the complexities of ecological care into lab work, transhumanist plots, and savior science—is not a nostalgic position but an argument for a different kind of futurity based on justice and care work across the human-animal-machine condition. Ultimately, the message de-extinction sends is that it may be easier to change the existential structure of animals than it is to do the work of careful and consensual environmental justice in conserving species, habitats, and the human communities that are long-standing stewards of ecosystems.

## Human De-Extinction Plots in Octavia Butler's *Xenogenesis* Trilogy

The paths of de-extinction technology intersect with recurrent science fiction plots of utopian and dystopian visions of biological enhancement. The biotechnological imaginary of transhumanism tends to accelerate evolutionary plots toward their extreme ends—either endless life or a sense of imminent extinction for life on a precarious and imperiled planet. Curt Meine remarks that the fantasy space of the resurrection and enhancement of species runs deep: "The prospect of de-extinction taps into deep layers of human myth and meaning, involving ancient themes of life and death, light and dark, creation and chaos, loss and resurrection, free will and fate, guilt and redemption."[37] Octavia Butler's *Xenogenesis* trilogy (completed between 1987 and 1989) situates ontological dilemmas concerning genetic manipulation, de-extinction, and the hybridization of life in scenarios that tie planetary futures with experimentation at narrative and existential levels.[38] Butler's novels design a storyworld in which the imagination of human survival is bound up with a prototypical science-fiction conceit of otherworldly visitation: instead of humans saving animals from extinction, an alien species swoops in to save humans, and also all of Earth's genetic life, from perishing.

While Butler's *Parable of the Sower* has become required reading in times of global warming and the rise of right-wing authoritarianism, her *Xenogenesis* novels have not received as much attention, yet they provide canny narratives on the malleability of biological life in extreme times. *Xenogenesis* gives an account of what it would look and feel like if the human species was the subject of resurrectionary genetic manipulation perpetrated by aliens who declare their motives to be biocentric. Instead of human scientists commanding de-extinction technologies over other species, Butler's extraterrestrials use their de-extinction methods to restore humans and repopulate the Earth, with some very questionable conditions attached. In Butler's telling, the remaining humans must negotiate new forms of compromised agency in a post-apocalyptic situation in which their own existential status hangs in the balance. The human characters in the novel must decide, when having one's own genome manipulated in a de-extinction event, what remains of the meaning of consent and cohabitation as the norms of being a species suddenly undergo change.

Elizabeth Hennessey, in a discussion of cryogenics and de-extinction, asks: "As scientists splice the genomes of different species in an attempt to bend back time, what kind of chimeric creatures might emerge from the cryo-archive?"[39] One imaginative and speculative response can be found in Butler's trilogy. I provide an extended reading of these novels to reflect on how Butler's writings fundamentally re-pose the question "What is extinction?" by placing de-extinction technology not in the hands of humans but in the control of intelligent alien animal-others who espouse biocentric values. Butler's hypothetical near future era in which de-extinction takes center stage finds that these salvational technologies do not resolve all ecological and existential crises but, instead, recapitulate the entanglement of biopolitical decisions over the metrics of species with valuations of life and death borne from such technologies.

The first novel, *Dawn*, begins a few centuries after nuclear warfare has decimated the planet, eradicated most terrestrial animals, and nearly eliminated humans. A small group of remaining humans have been captured and put into suspended animation by an alien species called Oankali who happen upon the devastated planet. The Oankali, tentacled and Medusa-like, have three distinct bodies, a male and female, and a third, non-gendered form called ooloi. Oankali need five bodies to mate, the male and female Oankali, an ooloi, and a male and female of another species. From the outset of the novel, we find that the Oankali, on their spaceship, which is also alive, have stored human DNA and begun manipulating it, learning to control human diseases and mitigate antisocial behavior and mood swings. Among the captive humans, the Oankali awaken Lilith Iyapo, a Black woman who they put in charge of awakening the rest of the humans, introducing them to the aliens, and planning their return to Earth. The few remaining humans are offered the choice to interbreed with the alien species and eventually leave Earth, or remain on Earth but be sterile, completing the extinction event. The Oankali announce themselves empathetically as gene traders and also are interested in making a genetic bank of the overall biodiversity of the Earth. Despite the mass extinction event of the nuclear war, "Earth was still a huge biological bank itself, balancing its own ecology with little Oankali help." By comparison, on the spaceship, "there was only a potential profusion stored in people's memories and in seed, cell, and gene-print banks" (481).

At several moments in the trilogy, the Oankali remind humans: "We have prints of all of you" (291), meaning that they have collected and stored the genetic material of humans as well as the plants and animals of the Earth. The Oankali themselves declare their manipulations of genetic material as a biological necessity: "We do what you would call genetic engineering. We know you had begun to do it yourselves a little, but it's foreign to you. We do it naturally. We *must* do it. It renews us, enables us to survive as an evolving species instead of specializing ourselves into extinction or stagnation" (40). Here, gene gathering and manipulation are described as being instrumentally necessary to the continuation-by-variation of the Oankali. The Oankali declare extinction to be as execrable as "stagnation," which implies that staying put in one species form is a biological impoverishment. The aliens seek to either persuade or impose this attitude upon the few remaining humans.[40]

My readings here discuss how the Oankali are motivated to archive the Earth's biodiversity but not save the Earth itself. The aliens are deeply curious about the Earth's animal life but have no deep connection to animals in particular, beyond the fact that their own spaceship is a giant, living entity. The Oankali do not see a contradiction in their desire to trade genes with humans and to store the Earth's biological material in seed and gene banks while they let life on Earth go extinct. The "salvage anthropology" practiced by the Oankali borders on extractivism and bioprospecting. The novel weaves human-animal-alien relationships so tightly that the sense of utopian possibility of cohabitation across species lines is bound up with dystopian appropriation of genetic diversity and human bodies that echoes with the history of human slavery and exploitation of animals.

There are already an extensive number of analyses of Butler's narrative in terms of genetic determinism, feminist science, and Afrofuturism. My reading builds on this scholarship while focusing on issues of speciation, de-extinction, and gene banking as central to the novel's vision of alternative biological pathways enmeshed with hypothetical extinction plots. In Butler's trilogy, ecological catastrophe and the use of gene banks are two sides of the same situation. As the Oankali explain to the surviving humans, the nuclear war had decimated the overwhelming majority of life on the planet and "had made their world utterly hostile to life" (290). Almost all the plant and animal life the humans see on the planet has been reconstructed from the "prints" of plant and animal genomes the Oankali had made. Without this re-engineering, "Nothing would have survived except bacteria, a few small land plants and animals, and some sea creatures. Most of the life that you see around you we reseeded from prints, from collected specimens from our own creations, and from altered remnants of things that had undergone benign changes before we found them" (290).

If intelligent humans have begun to preserve genetic material in seed and gene banks as a way to try to evade permanent ecological loss, one could argue that similar intelligent life elsewhere in the cosmos will have been confronted with a comparable problem and conceived of a version of genomic banking as well. It would appear that the Oankali demonstrate a distinctly biocentric ethos in their attention to archiving life on Earth and going to the trouble to "reseed" it once again for animal and human habitation. However, Butler's narrative, ultimately, offers a more ambivalent view of how the aliens use gene banks and reductionist gene science to "acquire new life" (41). As I will show, the Oankali do not revere the Earth's ecosystems, nor do they believe that the specific lives of animal species and their unique habitats are, ultimately, all that important.

The Oankali did store the Earth's genomic material, but when they reseeded the planet with life they mixed and matched the original genomes with other genomic material. The Oankali show they can recreate extinct animals but, just as well, create new animals by genetic manipulation. This example of de-extinction is quite similar to the mammoth 2.0 vision of current de-extinction efforts based on splicing genomes between different species. Another unsettling use of the gene bank in the novel arises from the issue that the Oankali's genome collection made from Earth's biodiversity is inseparable from their motivation to absorb human genes into their bodies, which, in essence, means effacing the unique distinctness of the human species even while preserving its genes.

Thus, in Butler's story, for human genes to survive, the specific human form and body must give way. The surviving human characters are faced with an existential dilemma whereby they (or rather their genes) can remain human but not alive, or alive but not human. Those who choose to keep the human body intact—but at the cost of fatefully eliminating the human genome since they are sterilized—are called resisters (in their view, the Oankali are implementing what could be called an "Oankaliocene"). Placing the onus for this decision on humans, the Oankali initially believed that the nuclear war demonstrated that humans had decided to commit "humanicide" (8). As one of the Oankali, Jdahya, tells Lilith, "We thought . . . that there had been a consensus among you, that you had agreed to die" (16). Yet, when the remaining humans are faced with self-knowledge of their own imminent event of extinction-by-metamorphosis, they are willing to use war against the aliens to keep the human form intact.

Readers of Butler's novel, who are meant, presumably, to identify with the humans, are caught in a dilemma similar to the predicament of ecological preservation efforts today. It matters that there continues to be the widest range of unique species forms that constitute diverse ecosystems, but one should not be too essentialist as to what constitutes a species. Species continue to evolve, and species forms can and will change through symbioses and adaptation. However, some sense of the essential uniqueness of a species form is necessary for extinction to matter at all. Otherwise, one could say that as long as there are complex genomes being reproduced somehow, specific species forms do not really matter.

It is common for conservation biologists today to declare the species "the basic unit of conservation"[41] but, at the same time, question what species and speciation really entails. In Butler's narrative, the human desire for being specifically and distinctly human is cast as either a form of resistance or nostalgia, as the Oankali force the human species to give way in a speciation event. Thus, this last human narrative intersects with a new, first human-animal-alien narrative, a human-Oankali hybrid. This first hybrid being, with the role of Adam/Eve played by Lilith, also stands as an allegory for the hybridizing work of the genre of science fiction. Xenogenesis is the generative name for the SF genre that invokes post-human figures to imagine an alternative to anthropocentric self-destructive tendencies.

The Oankali are fond of sampling and storing biodiversity, but are they what deep ecologists would call biocentric as counterposed to anthropocentric? As Nikanj, one of Lilith's Oankali partners, puts it, "We revere life" (153). For example, Akin, Lilith's human–Oankali child, takes every opportunity he can to "taste" the biological life around him, not to consume it but to store its genetic material while appreciating its presence. By contrast, "Humans had evolved from hierarchical life, dominating, often killing other life" (564). The Oankali define themselves as life made from other life and, thus, adhere to a kind of universal biological egalitarianism: "Oankali had evolved from acquisitive life, collecting and combining with other life. To kill was not simply wasteful to the Oankali. It was as unacceptable as slicing off their own healthy limbs. They fought only to save their lives and the lives of others" (564). In another example of this reverence for life, the Oankali do not eat animals and require that humans who mate with them do the same.

However, there remain stark tensions between the biocentrism of the Oankali, their gene storing and trading, and care for the specific biodiversity that the Earth holds. To care for biodiversity certainly involves stemming extinctions that would be not too hard to prevent or are caused by human negligence or self-interest. Yet the Oankali are not worried at all about the extinction of the specific form of the human species or, really, any other animal. The generic biocentrism of the Oankali is evident in their claim that "Life was treasure. The only treasure" (564). However, this phrasing also suggests that life in general is the only true value, not any specific life form. The Oankali are not concerned by the loss of Earth in the slightest, which they will leave a shell of itself by harvesting it of organic and inorganic material for their spaceships.

Another way of phrasing this difference between biocentrism and biodiversity is that the Oankali care about life (and diverse gene pools) but are not very attentive to specific animals themselves. The Oankali relish life at the level of genetic surplus and storage, but they are not very mindful of the accumulated deep-time history of evolution responsible for shaping each unique species form. Furthermore, the Oankali do not seem to have close relationships with companion species aside from their own ship, which is alive. One might say the Oankali revere genes and what can be done with them more than any actually existing animal.

While on the spaceship, Akin is drawn to a compositely assembled animal the Oankali had called a "tilio," which is described as being similar to an otter. "Assemble was the right word, Akin thought. The tilio had been fashioned from the combined genes of several animals. Humans put animals in cages or tied them to keep them from straying. Oankali simply bred animals who did not want to stray and who enjoyed doing what they were intended to do. They were also pleased to be rewarded with new sensations or pleasurable familiar sensations" (446–47). Akin responds with both affection and suspicion toward this constructed animal that seems to enjoy its captivity. He recognizes in this animal the tendency of the Oankali to treat all life as something amenable to control: "You controlled both animals and people by controlling their reproduction—controlling it absolutely. But perhaps Akin could learn something that would be of use to the resisters. And he liked animals" (447). Akin shows a care for the animal that is not the same as the Oankali attitude toward care for its genes and reproductive potential, which they seem to equate with "life."

This discussion of the dilemmas concerning extinction and de-extinction, gene banking, and reductionist gene science in "assembling" animals further complicates recent debates in literary criticism on whether or not Butler endorses a genetic essentialism in her novels. Jessie Stickgold-Sarah finds that all the characters and animals in the novel must, foremost, define themselves by their genetic profiles. Hence, for Stickgold-Sarah, "The gene figures in *Xenogenesis* in several ways, all of them distressing: as a tool of surveillance, as fate, as the threatened core of human identity, as a source of fundamental human flaws. Even when Lilith or other human beings challenge these interpretations, they are unable to advance any opposing metaphors."[42]

Hoda M. Zaki argues that Butler's acceptance of a biological hardwiring of human behavior is a politically conservative position, because it forecloses any notion that humans could act otherwise toward their own genetic constitution.[43] Sherryl Vint points out that it is really the Oankali who are the genetic essentialists, and that the humans in the story are not entirely convinced that their behavior is wholly structured by the "conflict" in their genes that they are told codes for hierarchy as well as intelligence.[44] Yet, as Vint also notes, the Oankali are largely portrayed as positive, while many of the humans act brutishly and refuse to learn from, even if not mate with, the aliens. Genetic essentialism, then, seems biopolitically opportunist for the Oankali and biopolitically fateful for humans.

In response to Zaki's argument, several critics have claimed, rather, that Butler's genetic essentialism is used strategically to critique other forms of oppressive essentialism typically attached to biology, especially racism. The Oankali do not mention race at all, and it has no biological bearing with them. Nancy Jesser comments that "Butler's essentialism should be read within a context of a gene theory that undermines racial categories and constructs, but that does not abandon genetic input in other human aspects, most importantly sex/gender."[45] Thus, for Jesser, a sense of genetic essentialism that treats genes as creative resources in common is needed to dismiss the pseudoscience of the hierarchical metrics of racial biology. Furthermore, Butler's narrative centralizing of genes shared across human–animal–alien life becomes the premise for human–alien sexual and reproductive acts that associate futurity with non-heteronormative and xenophilic models of filiation and symbioses.

However, in *Xenogenesis*, every form of reproduction is a contested, tense, insecure, and unstable activity. The Oankali need three differently gendered bodies to mate—a human male and female and an ooloi—and the Oankali acknowledge they are not sure how the "union" between the species will turn out. The humans who resist having hybrid children unwillingly condemn themselves to being non-reproductive (although later they are given the option to make a new colony on Mars that would allow for human reproduction). Reproductive futurity for humans in the trilogy turns into a critique of their own speciesism. For humans to reproduce and have a genetic future, they must open themselves to genetic otherness and forgo notions of species purity and the attitude that humans have dominion over all life.

Reproductivity as hybridizing with the alien other in this sense is non-heteronormative and non-anthroponormative. However, the layered reproductive decisions facing humans and aliens under complex conditions of duress and desire do not add up to a streamlined faith in the xenogenic future. Rather, sex and reproduction are caught up in intensely felt biological and psychological pressures that accrue around situations of life on the edge of extinction. In interviews collected in *Conversations with Octavia Butler*, Butler has noted that she wanted to imagine the Oankali as creatures who don't look like us yet desire us. Since they are more like amoeba than like mammals, Butler mentions, "I wanted to have the interesting task of figuring out exactly how a different form of sex might work biologically."[46] Her solution is to turn to reductionist gene science and imagine a kind of genetic consistency across species and exoplanetary worlds. This sense of a shared gene pool with aliens leads to fantasies of symbiotic and xenophilic futures as well as problems of biopower that arise in any act of manipulating the genes of humans and other animals, with or without their consent.

Whatever genes make up the Oankali, those genes themselves seek to combine with other genes in perpetual symbiosis. Oankali do not see genes as "selfish" (in Richard Dawkins' sense) in that they want to perpetuate only themselves. The Oankali would gladly trade genes toward continually becoming-other, rather than reproduce themselves and their current genes over and over again into "stagnation." Yet in another sense, the Oankali do participate in the discourse of "selfish genes," in that they see the gene as the primary locus of fate and character. Akin states to a human he tries to convince to go to Mars, "Human purpose isn't what you say it is or what I say it is. It's what your biology says it is—what your genes say it is" (501). Genes seem to do all the speaking for the Oankali.

Evelyn Fox Keller has written extensively on the view, so dominant throughout the twentieth century, whereby the gene is said to be the origin and primary site of activity that defines the organism. Keller provides an intricate history of how the gene became such a singular obsession for biologists, but also indicates how much of recent biotechnology is moving away from such a narrow focus on genes only and toward more interactive models of developmental biology, ecology, and epigenetics. Keller adds that the shift away from the gene as lone actor is due not only to better genetic science but also a better understanding of how science is connected to institutions and social practices. For, as Keller remarks, "What is specifically eclipsed in the discourse of gene action is the cytoplasmic body, marked simultaneously by gender, by international conflict, and by disciplinary politics."[47] This interactional approach to the gene includes recognizing how genetic knowledge is implicated in contested social values and recent historical events of the enthusiastic embrace of genetic manipulation, which brings us back to the world of the Oankali.

Ultimately, reverence for life, as declared by the Oankali, means storing, manipulating, and mixing genes rather than caring noninvasively for the unique lives and habitats of specific animals (and one can imagine a biotechnologist today saying the same thing about revering life while aiming to manufacture all sorts of new hybrid organisms, such as a mammoth 2.0). Akin comes to fight for the restoration of human autonomy not just because he is part human but also because he is wary of how the Oankali treat genes and animals generally. Akin is skeptical about some of the biopolitical practices of the Oankali even while he recognizes the fatal "conflict" in the human genome (as the book calls it) that codes for both hierarchy and intelligence and is blamed for leading humans to humanicide. Forcing humans to trade genes with the Oankali will apparently resolve the genetic contradiction that led humans to kill themselves and other animals. But Akin recognizes that the Oankali would also be responsible for the ultimate extinction of humans: "Because of us, they won't exist anymore" (377). After realizing the role the Oankali are playing in human extinction, Akin exclaims, "There should be Humans who don't change or die—Humans to go on if the . . . unions fail" (378).

Akin's insistence that the specific form of the human not go extinct but be allowed to live on Mars is consistent with how other humans critique the Oankali for treating them derisively as animals lacking self-determination. That is, the Oankali treat humans the way humans tend to treat animals, as instrumental means for human ends. Lilith remarks on this comparison early on in the first novel after hearing of how the Oankali have been learning about human bodies by manipulating their DNA while they sleep: "We used to treat animals that way . . . We did things to them—inoculations, surgery, isolation—all for their own good. We wanted them healthy and protected—sometimes so we could eat them later" (33).

Sherryl Vint calls narratives about alien treatment of humans as animals "reversal stories."[48] Vint describes this narrative conceit as a device to show a "perceptual world and cultural norms [that] are different from one's own" (158). Humans who treat animals reductively as mere genetic matter or means to satisfy human desires are themselves the object of reductive treatment by the Oankali. The aliens are effectively establishing their own frozen zoo of human DNA, reversing the collector/collection relationship of today's frozen zoos. "They collect the eggs, store them, collect sperm, store it" (295), as one Oankali matter-of-factly notes. Akin balks at this zoo-like treatment, and the version of reductionist science implied by such banking and manipulating, while bonding with other animals. In effect, Akin seeks to reverse the reversal story and learn from this situation by preventing unnecessary human extinction and caring about the specificity of animal lives without regard to prospecting for genetic difference.

Reversal stories connect to "becoming-animal" stories in the lexicon of Gilles Deleuze, but we should be careful here in assessing how much these two motifs overlap in the context of the discussion of extinction and reductionist gene science. While Butler's trilogy beckons the reader with the fantasy of becoming-alien, a version of becoming-animal, the narrative poses hard questions about extinction and the loss of biodiversity that are not addressed in philosophies of becoming. Becoming-animal in Deleuze's thought is not an end in itself that demonstrates solidarity with biodiversity per se but a stepping-stone to a more powerful ontology of continuous individuation. Deleuze claims that individuation entails a rudimentary vitalist power called "a life" that is inextinguishable in the universe because it flows from the immanent forces of the universe interacting with itself. This vitalist power is ever-changing and never can be vanquished. Deleuze calls "a life" "an impersonal and yet singular life."[49] "A life is everywhere . . . an immanent life carrying with it the events or singularities that are merely actualized in subjects and objects. This indefinite life does not itself have moments . . . but only between-times, between moments" (29). Such impersonal life transects the life and death of individual bodies and species.

This vitalist view of a perpetual "indefinite life" is not strictly biological and never quite explained in physical-evolutionary terms but follows from inexhaustible forces of difference, repetition, and affirmation of multiplicity and becoming. These immanent powers and forces are implicit in all beings, yet "a life" particularly manifests in intense moments of individuation, such as when an organism passes into a new state of becoming or dies. This death is, however, only a temporary pause or "playing with death" (28) in the continual becoming of "a life" that will carry on in another form. Ultimately, then, by Deleuze's accounting, extinction is only a local and temporary phenomenon, as this virtual and vitalist life cannot go dormant and "might do without any individuality" (30).

In Butler's trilogy, becoming-alien means absorption into a continuous effort of genetic and morphological change that discards species specificity for the sake of ongoing life. One of the ooloi constructs named Ahajas tells a human:

> If I died on a lifeless world, a world that could sustain some form of life if it were tenacious enough, organelles within each cell of my body would survive and evolve. In perhaps a thousand million years, that world would be as full of life as this one. . . .
>
> Nothing is more tenacious than the life we are made of. A world of life from apparent death, from dissolution. That's what we believe in. (662–63)

Life without death approaches "a life," which Deleuze believed would be "complete bliss."[50] Yet as Ahajas nears a state of crisis over his need to find a mate in the final novel of the trilogy, its body starts to lose its form and threatens to turn into a less complex arrangement of life that would be deemed a permanent loss of its previous identity. Ahajas begins to panic and viscerally resists such an outcome, seeing this possibility as an unwanted evolution, even if not a death. In fact, the Oankali cannot really promise a world free from dissolution, even if death is a non-factor, as its body would effectively sprout life anew in some way. Without a mate, ooloi risk losing their own biological specificity to the extent of becoming a kind of organic soup. The ooloi and Oankali, just like humans, strive to resist a certain loss of form, memory, and physical complexity that would set their bodies into a gelatinous state.

Becoming-alien, even as it undermines the human/other hierarchal distinction, is not necessarily compatible with care for the nuances of biodiversity and the uniqueness of animal lives. Butler's story shows that the seemingly utopian vision of continual biological change and bodily modification—which also is the ultimate horizon for transhumanist existential technologies—may not be the ethical means by which to leave behind the impasses and oppressions of the category of the human. Instead, new impasses and oppressions arise when the body is treated as storable and continually changeable, and the terms of biodiversity and ecology become subsumed into genetic modification and the biopolitics of control and consent over life.

Neither gene fetishism nor continual becoming are adequate to attend to the nuances of the species form situated in place- and time-specific ecosystems that are susceptible to the devastation of extinction. The Anthropocene is the time when humans have extended control of the Earth to become a force of geological proportions, yet it is also the time in which humans find themselves questioning such power and its repercussions for other life on the planet. Amassing frozen zoos and experimenting with ways to circumvent extinction through techno-science are projects—which we can now define as acts of xenogenesis—that can expand the scope of responsibility for animal futurity. But the discourses and practices of animal care used in these cases is not justification alone and inevitably will disrupt the definitions of biodiversity and care that they presuppose.

De-extinction science is an existential technology that fundamentally redefines extinction—potentially redefining all of animal existentiality from birth to life to death—so the rationales for this science as serving as a corrective to past extinction wrongs are undercut by changing the meaning of species finitude. Gene banks and de-extinction, as well as revivification and hybridization, are complexly entangled phenomena that always involve species in the most fundamental questions of power over life and death, the fundamental coordinates of our past and future allegories. Confronting the biopolitics of de-extinction technologies and scenarios does not mean simply taking the side of "life" in contrast to technological forces—siding with the abstraction "life" in whatever form it takes is already fraught, as Butler shows—but, rather, approaching the specific lives and species of animals with a sense of care that is not dependent on expensive and salvational investments.

Contemporary theorizations of ecology, entanglement, and precarity are replacing older intellectual paradigms of "life itself" or generic existentiality that were developed to counter the narrow determinism of Darwin-era species talk. Butler's novels invite us into this more self-aware ecological transition with reversal stories, speculative plots of genetic science, and alien genres at the tense intersection where multiple extinction narratives converge in scenes of last human and animal conditions. Cary Wolfe points out how the rapid expansion of our current biopolitical framework "increasingly takes as its political object planetary life itself."[51] Butler also understood how the predicaments of the life and death of species might need to be posed at an interplanetary level, as well, to properly comprehend the stakes of biodiversity on Earth.

# Conclusion

There is a recurring argument made among advocates of outer space exploration that humans must find a way to leave Earth as a way to minimize, if not outright escape, the risks of extinction that all terrestrial animals face. If humans stay on Earth, they certainly will become extinct eventually, as the planet has an expiry date. The increasing luminosity from the sun will sterilize the earth and eradicate all life in about 3.5 billion years, but most complex life will have perished long before then, in the range of 0.9 to 1.5 billion years from now.[1]

However, if humans—or perhaps our post-human descendants—become a multiplanetary species, then it may be possible to outrun extinction for a very long time. At the very least, colonizing other planets appears to be imperative for human continuation because it increases the chances of survival of the species if something bad happens to life on Earth. The lesson extracted from the demise of dinosaurs due to an asteroid strike is that a species needs to have more than one planetary home. Carl Sagan viewed space exploration as inevitable for any advanced civilization because life on just one planet would be too risky:

> Since, in the long run, every planetary society will be endangered by impacts from space, every surviving civilization is obliged to become spacefaring—not because of exploratory or romantic zeal, but for the most practical reason imaginable: staying alive. And once you're out there in space for centuries and millennia, moving little worlds around and engineering planets, your species has been pried loose from its

cradle. If they exist, many other civilizations will eventually venture far from home.[2]

Stephen Hawking has urged humans into spaceflight for similar reasons. "The human race shouldn't have all its eggs in one basket, or on one planet," Hawking states.[3] Like Sagan, Hawking foresees that leaving Earth is a necessary matter of survival: "Our only chance of long-term survival is not to remain inward looking on planet Earth, but to spread out into space." Elon Musk finds that colonizing other planets should be done in the name of a devotion to humanity: "I think there is a strong humanitarian argument for making life multi-planetary, in order to safeguard the existence of humanity in the event that something catastrophic were to happen. . . . I think we have a duty to maintain the light of consciousness, to make sure it continues into the future."[4]

For Musk, the foremost humanitarian act is keeping humanity in existence. However, the "light of consciousness" is not the same as human life (or an ecological life), and many in the artificial intelligence community support a vision in which the future development of intelligence need not be tied to a human body. Nick Bostrom advocates for an enhanced humanity to become eventually transhuman (either as a human-machine or a fully artificial intelligent entity) to pursue "astronomical" potentials for realizing "value" of the greatest amount of good for the greatest possible number of happy and intelligent lives.[5] For this astronomical value to be realized, transhumanity certainly would have to depart Earth and explore the galaxy in search of resources and other intelligent life.

Every venture into space has the potential to change the stakes of extinction. Mars could be Planet B. Colonizing Mars and terraforming it would require transplanting much of Earth's biodiversity. Being extinct on Earth would not, then, constitute being extinct as such. After becoming multiplanetary, humans would seek to become multi-stellar inhabitants. In these scenarios, space colonization can be seen as an extension of the Enlightenment project as the pursuit of a rational collective endeavor as far as possible in time and space. Yet, as discussed in the introduction, Enlightenment reasoning also entails the recognition of the ecological and conceptual ramifications of finitude and the need to develop modes of collective care, self-critique, and mourning as ways of acknowledging shared precarity. Recognizing extinction on Earth should not be reductively treated by abandoning terrestrial life; and, as Darwin recognized, species finitude can, paradoxically, contribute to species plurality. Finitude both expands and contracts the potential space of existences. Still, for the most part, one must reckon with the reality of extinction today as not regenerative but annihilative.

Perhaps spacefaring will not become a fail-safe way to improve the chances of the deep-time survival of humanity and Earth-based life. Stephen Hawking also warned that humanity should not seek or want to come into contact with intelligent alien life. Hawking argued that if there is even the slightest possibility such aliens could be hostile, human life, and perhaps all of Earth's life, would be in grave danger. Extraterrestrial intelligent (ETI) life would likely be much more technologically advanced and, thus, capable of great harm if there was something of interest for the taking on Earth. A disastrous encounter with alien life also may happen even if those aliens have no superior intelligence. Kim Stanley Robinson, in his novel *Aurora*, writes of a generation ship sent to colonize a nearby planet.[6] Shortly after arrival, the humans discover the planet is populated with a deadly virus unlike those found on Earth and without any possible means of being immunized. Leaving Earth, thus, could be the one thing that most jeopardizes life on Earth.

In the case of terraforming Mars as a plan to provide a backup for life if something goes wrong on Earth, it is hard to imagine that humans will be able to leave behind their earthly problems while on Mars. Most likely, the same ecological problems and social conflicts currently besetting life on Earth would follow humans to this "off world." Would weapons be banned on Mars? Would humans be better at sharing on Mars and more conscientious of extinction? Or would Mars become another kind of moral hazard, such that people would be more likely to disregard the welfare and work of sharing a planet if they knew there were alternatives or backup plans?

Settling on Mars would require immense extracting and harvesting of resources on Earth and a similar extractivist project on the new planet. Such resource gathering would exacerbate and disproportionally affect already marginalized and racially-marked human communities whose lives are oppressed by the drive to harvest these materials. Extractivism also marks "the end of nature" for the biodiverse life located in these resource-rich zones; for example, the orangutan of Borneo whose forests are being clear-cut and replaced by palm oil plantations (IUCN data indicates 50 percent of the orangutan population has been killed in the past forty years[7]). Intensified extractivism on Earth, even if done in the name of "saving" such life, would accelerate extinction. The efforts to colonize Mars likely will proceed in neglect of the need to do the work of decolonizing and fostering environmental justice on Earth. And, finally, asteroids, gamma rays, and the sun's inevitable expansion also afflict Mars. All these same problems arise on any other planets and in other solar systems we might be able to reach. And we still would need to, once again, develop new modes of planetary sharing that characterize our Earthly sense of "world," no matter where we go. The idea that only by leaving Earth might we be able to save it presents another contradiction implicated in the conditions of extinction.

Becoming interplanetary is still a worthy goal for reasons scientific, communitarian, and aesthetic—and, hopefully, for diplomatic reasons if contact is made with another intelligent life form—but space exploration most likely does not reduce extinction risks. Wherever we travel, we will take our extinction problems and existential risks with us. And space exploration will have costs to biodiversity on Earth, if only because the rush to spend trillions of dollars in efforts to explore other planets and make contact with ETIs stands in stark contrast to the rising extinction rates devastating species at home. The same mentality—and the same technology—that allows us to leave the Earth in the name of securing life also puts all life as we know it at risk. To be on Earth is to be imbricated in the dilemmas of sharing of the Earth. Will interstellar travel be an event of sharing the Earth with others, or will it be a "great unsharing," an abandonment or endangerment of the Earth?

Understanding extinction necessitates close and critical analysis of what it means to call something "the last" or invoke a sense of conclusiveness or finiteness of a species, a form, or an event. This book has provided an examination of key scenes of extinction and human and animal lastness over the past few centuries. The pursuit of firstness and lastness in space—the "final frontier"—will yet again redefine what life means and what extinction means.

In contrast to the evocations of outer space as the last frontier beckoning first explorers, I want to examine a very different kind of "last" venture into space. The artist Trevor Paglen, in 2012, obtained permission from a satellite company to affix a disc on the side of a satellite set to orbit Earth.[8] Paglen initially conceived of an art project called *The Last Pictures* to send 100 photographs into space. At first, he considered this to be just conceptual art but then found the satellite maker Echostar Communications amenable to bringing the idea to fruition. Paglen and his team selected the photographs and had the images etched onto a silicon plate slightly over 2.5 inches in diameter, which was sealed inside a second cover made of gold with a star chart showing the position of the solar system in the Milky Way. The disc, built to remain intact for over a billion years, is attached to the satellite *Echostar XVI*, which transmits radio waves for high-definition television and phone signals for North America and has a battery lifespan of about twenty years (see figures 7-1 and 7-2). After that point, the satellite will be decommissioned and pushed further out to orbit into an area that has become a junkyard for power-less space equipment. The dormant machine, then, will be left to orbit the earth for a few billion years.

The implantation of a disc on a satellite as a message awaiting discov-ery by an extraterrestrial is an homage to Sagan's own famed gold discs affixed to the *Voyager* satellites launched in 1977. Sagan's team etched onto the records almost ninety minutes of musical selections (ranging from Bach to Senegalese percussion to Louis Armstrong), encoded photographs, an audio essay, and a series of greetings in fifty-five lan-guages. Each satellite included an attached stylus with indications on how to play the disc. The record became a new kind of message that com-bined an archival representation of cultural diversity on Earth and a future greeting to whomever might find the satellite.

Sagan's team also arranged for 118 photographs to be encoded in ana-log fashion in the grooves on the record and provided instructions for to how to translate the information back into images. In selecting the photo-graphs, the team ruled out images of war, crime, and poverty, concerned that such images would be interpreted as conveying hostile intentions. The range of themes covered by the images include the human body and stages of life, pleasant images of nature, "Family of Man"-style displays of human diversity, everyday human activities, and demonstrations of industrial and technological prowess. The archive is a slice of 1970s technological optimism and social conviviality, offering itself as an example of scien-tific and humanist cultural convergence, a suggestive formula for eschew-ing conflict and modeling the need for imaginative collaboration as a planetary whole.

The discs presented a special opportunity to say something about human life to any alien who might stumble upon the satellites. This ar-chive also became a moment to consider what to say about being human on an object that likely would well outlast humanity. Reflecting on the voyages of these discs, Sagan pondered how the message might be received by an extraterrestrial encountering the records some tens of thousands or more years from now. The ETI would surely wonder, "Had we destroyed ourselves or had we gone on to greater things?"[9] The unknown future fate of these *Voyager* discs, carrying messages that will reach at least one intelligent species (ourselves), helps frame thinking about extinction in present times. Sagan felt confident that the launching of a satellite with a

FIGURE 7-1. EchoStar XVI satellite, photograph, 2012, copyright EchoStar
Corporation, Courtesy Space Systems Loral.

FIGURE 7-2. The Last Pictures Artifact, 2013, Golden Disc, photograph, copyright Trevor Paglen, Courtesy of the Artist, Metro Pictures, New York and Altman Siegel, San Francisco.

message of greetings and unity-in-diversity would bode well for our own planet and any possible receiver of such message. "But one thing would be clear about us: no one sends such a message on such a journey, to other worlds and other beings, without a positive passion for the future. For all the possible vagaries of the message, they could be sure that we were a species endowed with hope and perseverance, at least a little intelligence, substantial generosity and a palpable zest to make contact with the cosmos" (236).

Since the *Voyager* gold discs, there have been many objects and radio transmissions sent into space to broadcast a potential legacy for humans. These missives form a speculative archive that may never be encountered. They carry within the techno-existential aspirations of a moment in time on Earth as well as a certain way of reflecting on extinction in the present. Paglen's photographic archive stands as a counter-allegory to Sagan's disc in its refusal of self-censorship and inclusion of a heavy dose of self-critique regarding the reasons for sending such missives into space. The images selected for the *Echostar XVI* cover a much wider range of human

FIGURE 7-3. Carlton E. Watkins, Boston hydraulic mine, Nevada County, California, photograph, 1870s, Bancroft Library, University of California, Berkeley.

behavior than the *Voyager* archives, and do not shy away from showing disturbing and violent images in juxtaposition with beautiful ones. There is a picture of a human feeding an orca indicating human–animal conviviality, while another image is of a cramped battery of cages confining egg-laying chickens. A photograph of an avant-garde work of sleek skyscraper architecture contrasts with a photo of hydraulic strip mining taken by the well-known early California photographer Carlton Watkins (see figure 7-3).

There are several images of massive-scale infrastructures, but these no longer read so evidently as feats of technological prowess in the way they did for Sagan. Instead, each image raises questions about the darker implications of development. An image of the Hoover Dam being built resonates as the launching of the era of the mega-dam, regional fights over water, and dispossession of Indigenous lands in the name of national need. A photograph of what could be anywhere suburbia is actually the planned city of Levittown, New York, which initially sought to institute a policy that only whites could be homeowners. The photograph of a shipyard in the Baltic Sea (see figure 7-4) turns out to be previously classified govern-

FIGURE 7-4. Nikolaev shipyard, Black Sea, photograph, 1983, Courtesy of the artist and Creative Time.

ment material: the image was made by a U.S. spy satellite watching Soviet military buildup and shows an aircraft carrier. This photograph was leaked to the press by Samuel Loring Morison, who, subsequently, was prosecuted and imprisoned under the Espionage Act.

Alongside photographs displaying technological triumphs such as disease vaccinations, supercomputer development, and space missions, other selected images of infrastructure and technology are carceral (Illinois State Penitentiary), used for surveillance (cameras watching migrants crossing the United States-Mexico border), and enacting discriminatory state policies (concrete wall in the West Bank). Photographs of a pastoral nature are juxtaposed with images of climate change, including melting glaciers and nuclear detonation tests in the Bikini atoll.

The first image chosen is the back of the paper drawing of Paul Klee's *Angelus Novus* (1920), interpreted by Walter Benjamin as the angel of history who views a pile of wreckage accumulating along the path of historical progress as World War II approaches. According to Benjamin, the angel, its back turned against the future, has its wings stuck open by a storm blowing from Paradise. "What we call progress is *this* storm."[10] The angel, neither human nor animal but having elements of both, sees also the increasing loss of life across all species implicated in the perilous dramas of these historical events. The historical and meteorological references of Benjamin's reading of the image are repurposed in Paglen's use, as the satellite and its archive are positioned to watch over the Earth in an era of global heating and rising extinction rates, having no power to intervene in history but only to testify as a witness to it all.

Paglen's collection does not represent a great cumulative image of humanity, which distances this project from the *Voyager* discs that sought to convey a happy human family. The images on *Echostar XVI* are not aimed at any single characterization or summative concept of the human. As one cycles through the images, no clear narrative emerges—we see a mixed collection conveying exhilaration, junk, violence, ecstasy, animal tenderness and exploitation, super-technologies, planetary cruelties, insider cultural references, and cosmological longings. In contrast to the great big "hello" of the *Voyager* disc, this archive says, "I have no idea who we are." There is a refusal to monumentalize or self-aggrandize at the species level—yet, at the same time, this is an archive effectively forever.

Paglen conceived of *The Last Pictures* as embodying "the Anthropocene contradiction between the hyperspeed of capital and the deep time of anthropogeomorphology, the torrential flow of twenty-first-century pictures and their utter ephemerality."[11] Paglen compares the "forever" of this archive to the deep-time longevity of plastic trash and the electromagnetic radio waves emitted from commercial television radiating out for eons into space: "A cup of fast-food coffee is meant to be sipped for a few minutes, but its Styrofoam takes more than a million years to biodegrade" (xii). The Styrofoam cup is intended to be used in a blip of time, but its epochal material durability seems to mock human striving for permanence. Like the "fossil media" of the Styrofoam cup, the coming ruin of the *Echostar XVI* takes its turn within "the fractures and folds in spacetime that form where the timescales of economics and politics collide with the deep time of the Earth" (xii–xiii). Paglen's project presents the death drive of the Styrofoam cup and the communications satellite—objects that exemplify the compulsion to persist but also stand as husks of abandoned social relationships and discarded desires—in connection to the death drive of humans, framing our aggressive commitment to transform the planet into an image of our own self-preservation that coincides with an image of our negation.

Instead of monumentalizing humanity, *The Last Pictures* functions as an ultimate archive that also is a reluctant archive, a heap of broken images riding on a soon broken-down and powerless machine. No matter what they show in terms of content, at the level of form, these photographs can be considered another example of the "extinction shot." In each iteration of the extinction shot, we witness and yet question the ability to see at the same time. What these images show, in their being let loose from Earth, is a moment of "that-has-been" on Earth. These are now images of futurelessness as much as they are an address to future visitors. The *Echostar XVI* is both a testament to the existential reality of life today on Earth and a hurtling piece of junk once it is powered down, a combination of profundity and pointlessness. Paglen's work is a double allegory for the Anthropocene condition and the longevity of art.

No artwork is immortal, yet all art strives for a taste of the transcendental; both the durability as well as destroyability of an artwork are fundamental to the conditions of possibility of any art. Both the endurance and ruin of a work operate simultaneously in structuring the existential aesthetics of the work's form. A similar reflection is required for thinking about biological extinction and how the forms we use to cognize about extinction are themselves caught up in conditions of finitude. Methodologies developed by the avant-garde in engaging the limit ends of art, thus, intertwine with thinking about extinction as the limit ends of biological and cultural forms.

Hannah Arendt highlights the durability of art as a special kind of longevity that could provide a stable frame for the human condition and the desire to build lasting worlds. "It is as though worldly stability had become transparent in the permanence of art, so that a premonition of immortality, not the immortality of the soul or of life but something immortal achieved by mortal hands, has become tangibly present, to shine and be seen, to sound and be heard, to speak and be read."[12] The durability of art mediates between mortal hands and "a premonition of immortality" that emanates from human thought and desire. But Arendt notes that "this reification and materialization . . . is always paid for, and that the price is life itself: it is always the 'dead letter' in which the 'living spirit' must survive, a deadness from which it can be rescued only when the dead letter comes again into contact with a life willing to resurrect it, although this resurrection of the dead shares with all living things that it, too, will die again" (169). Arendt draws out a materialist and formalist existentiality of the art object. The potential perishability and readability of the work of art go together—the work can only be possibly "alive" and encountered afresh because it can also be "dead," perhaps never to be encountered again.

Jacques Derrida, in *Archive Fever*, discusses how archives can be both a storehouse for memory and a recognition of the destroyability of all the efforts and materials that inscribe and collect such memories. "There would indeed be no archive desire without the radical finitude, without the possibility of forgetfulness."[13] In the archive, memory and forgetting, *arche* and *telos*, first and last conceptualizations converge. Paglen's archive is a repository for first addresses (first greetings, first contact) and last addresses (memorials and mournings of Earth). Recall that, as described in the introduction, the first naturalists to discern the scientific fact of extinction grasped the entire Earth as an archive, seeking to make legible the deep-time historical record of geological and biological activity. In reading this record of fossilized finitude, the naturalists also came to understand how "the archive should call into question the coming of the future,"[14] as they realized themselves to be implicated in the pitiless work of extinction.

Arendt thought of durability as crucial to world making and world sharing, and she praised the enduring evidence of human craft, works of art, the great statements and deeds of humankind that live on in reputation, and long-standing institutions of law. But conditions of durability have changed when talking of plastic cups retaining material form for over a million years, or a satellite existing for a billion years. Durability also is a curse and a matter for extinction in the sense that the persistence of such objects can become pollutants noxious to ecosystems for many generations and will likely long outlast humanity. Paglen writes: "From the moment I began thinking about dead satellites and suicidal civilizations, the ghost machines in Earth orbit become an allegory for what happened to the people responsible for them."[15] Whoever might stumble upon this satellite archive, most likely in the far-distant future, will have to come to terms with either why the makers of the object disappeared or by what means did such people persist. As Paglen puts it: "In the future, we are the ancient aliens" (7).

This book's conclusion is itself caught up in the questions and responsibilities of last thoughts. How does the end of an address or a conclusive statement contribute to the organizing principle of thinking about the disappearance of a species? I have preferred to structure this book around a question repeatedly posed in different natural and cultural contexts and time periods—What is extinction?—rather than posit a definitive conclusion to this question. Phrasing extinction as a question means being dedicated to examining the changing meanings of extinction, the "periodizations" or historically specific moments in the redefinition and intensification of extinction. Posing extinction as a question leaves open the matter of future definitions and species entanglements.

Endangered life presents an imperative for acting with immediate care, but there is no single way, and no formula, for how to respond to extinction. It is possible to recognize that our definitions and foundations for care are being put into question by extinction and yet still commit to an ethics of species and bioconservation. It also is possible to reject biotechnological attempts at "solving" extinction or preserving all life while still supporting biotechnological contributions to conservation. Understanding the ecology of extinction—as biological phenomena and existential condition—does not require committing to what Hans Jonas calls the "unwanted, built-in, automatic utopianism"[16] of biotechnologies of life extension in pursuit of, as Ray Kurzweil states, the goal to "live long enough to live forever." Being committed to thinking about extinction and caring for the contingencies of species in environmentally distressed times involves both conservation and re-invention, attending to future transformations and retrospective responsibilities regarding the multiple definitions and answers in response to the question "What is extinction?"

Extinction is so extreme and demands our utmost attention, but it would not be possible to endure such thoughts if extinction came to be permanently in the foreground as the utmost concern of everyday life. It would be impossible and exhausting to think about and respond to all extinction threats for all species at all times and places. The discourse of extinction is so urgent and tremendous, as it should be, that it cannot be registered and comprehended as a continual daily phenomenon or sustained as a permanent emergency. Extinction potentially puts everything we know about nature and culture into a crisis so radical that the devastation of worlds becomes the only thought left to think. Yet to always and unrelentingly think at the limit ends of thought would devastate thought. The demand for constant vigilance and response would incapacitate the ability to respond properly with care.

At the same time, extinction cannot be cast into the background, referred to only when a species nears perilously low numbers. It is important to be haunted but not be so devastated that all we can see are ghosts. Yet, as Eileen Crist comments, most people rarely think about extinction at all: "Amnesia about the living world and its diverse beings is the wretched existential condition humanity has obtained in exchange for domination. A pinnacle of alienation from the natural world in our time has been the perverse public invisibility of the mass extinction episode on the immediate horizon. Human-driven mass extinction remains publicly largely unknown, little understood, rarely talked about, or summarily glazed over with platitudes."[17] Refusing this amnesia means attending to the demands of thinking about extinction and caring for species, and also caring for those who do the daily care work for species. There is a huge burden in economic costs and affective labor in doing the work of bioconservation. Double attention is needed to contest the exhaustion of animals and the exhaustion of caregivers who provide everyday support to endangered animals.

To think about and engage with extinction entails understanding how extinction affects all life, and also how some bodies and lives and species are more exposed to immediate existential threats than others. Events of extinction over the past few centuries have been cause for acts of solidarity among humans and non-human animals in some cases, while other cases involve the destruction of appeals to species commonality. Reckoning with finitude can lead to human exceptionalism (asserting that humans are the only species to understand and act upon their mortality), or it can undo human exceptionalism by revealing a precarious life in common with other species. Thinking about extinction can be what prompts some people to want to leave Earth behind altogether, while it also can cause people to bond closer with the ecological condition of a shared Earth. Sharing the Earth entails supporting long-standing relationships to bioregions among Indigenous peoples, who are the enduring land and species protectors. It means dismantling extractivist colonialisms and contributing to Indigenous and postcolonial resurgences. It means practicing new ways of grieving and loving together in a way that does not exhaust life. It means being committed to expanding the space of diverse existences on Earth (and perhaps beyond the planet). It means being open to diverse forms of kinship and nurture across nature-cultures. It means developing technologies of care and carefulness, technologies in support of environmental justice, social justice, and anticolonial work, to benefit the diversity of life on Earth and the diverse caregivers and custodians of life.

In the decade spanning 2010 to 2020, the IUCN listed 171 species that researchers confirmed as extinct (several of these species had passed away in previous decades but were deemed "data deficient" and required further population surveys to confirm extinction during the past decade). Species determined to have perished in the past decade include the Alaotra grebe, the Pinta tortoise, the splendid poison frog, the Bramble Cay melomys, the Western Black Rhino, the Chinese paddlefish, and the Catarina pupfish.

The Catarina pupfish, a freshwater fish, existed only in one river spring in Mexico whose water dried up due to increasing local aridity and heavy use for agriculture. Attempts to reproduce the fish in captivity did not succeed. The extinction of the Catarina pupfish, like all the species discussed in this book, is an event in nature and culture, and the term "nature-culture" indicates not a seamless fusion but mutual entanglement in situations that can be both enabling and destabilizing for making multispecies worlds in common. The loss of the Catarina pupfish, as with the extinction of any animal, forces us to ask how we can care and share better at the limits of life.

As a traumatic limit to nature and culture, extinction is the counterpoint to idealizations of nature as comforting "home" or primordial nurturer, but this does not mean nature is inevitably destructive. Extinction as a biological fact does not translate into a metaphysical principle that nature "wants" to destroy us. To pose the question "What is extinction?" today requires understanding how the demise of a species is caught up in biopolitical choices over who lives and dies and definitions of "biovalue" favoring some lives over others. But events of extinction also point to forms of biocultural sharing that involve collective and cross-species experiences of loss as well as attachment, with grief and desire together leading to new kinds of nature-cultures. Understanding extinction means thinking existentially about the many forms of multispecies beings with which we share the planet, and it means thinking about the multiple ways of cognizing and affectively responding to the end of a unique way of life. Becoming knowledgeable about extinction also means cultivating thinking at its limits and examining how conceptual limits intersect with natural-cultural entanglements that affect the entirety of life on Earth. Extinction is both something that requires interpretation and is the end of interpretation. As I hope to have shown, understanding extinction means becoming rededicated to sharing the Earth and, even in a time of animal lastness, developing new forms of address and redress across species.

# ACKNOWLEDGMENTS

Becoming knowledgeable about extinction, for me, has meant thinking about ecology and thinking ecologically. Thinking ecologically is about the forms of thinking as much as the content, and emphasizes reciprocity, generosity, thankfulness, connection-making, reparativity, and sharing. Such thoughts are closely connected to acts of care and gratitude. I cherish this moment to have the chance to thank the many people and places that have helped and supported me as I worked on this book. I gratefully acknowledge writing this book at Western University, situated on the traditional lands of the Anishinaabeg, Haudenosaunee, Lūnaapéewak, and Chonnonton peoples, a place that continues to be home to diverse Indigenous communities that are long-standing stewards of the land.

I have had the honor to present material at length from this book on special visits. These experiences have meant a great deal to me. Brett Buchanan invited me to deliver some initial ideas at his outstanding conference "Thinking Extinction" at Laurentian University in 2013 (where I, also, surprisingly, found myself sharing a scotch with Graeme Gibson and Margaret Atwood and talking with them about extinction photography). I deeply appreciate the opportunity to have been invited to present some of this book's material at visits to Slought Networks in Philadelphia, UC Riverside, Queen's University, and SUNY Buffalo. Thank you to Aaron Levy, Jean-Michel Rabaté, Sherryl Vint, Glenn Willmott, Judith Goldman, and Damian Keane for welcoming me and for your very gracious invitations. Special thanks to Glenn for his support when I needed feedback on this manuscript at a crucial juncture. I also delivered several works in progress at the biannual conference of the Association for Literature, Environment, and Culture in Canada (ALECC), as well as conferences run by MLA, ACLA, SLSA, and ASLE. Sébastien Lefait and LERMA at Aix-Marseille Université generously welcomed me to present material as I was finishing the book. I am grateful for the many thoughtful questions I received from the audience during these visits.

At Western University, I am deeply thankful to teach and learn with wonderful colleagues. I am especially appreciative for conversations on this book with Bryce Traister, Thy Phu, Jonathan Boulter, Matthew Rowlinson, Pauline Wakeham, Allan Pero, Manina Jones, Tilottama Rajan, Kim Solga, Julia Emberley, Jan Plug, Madeline Bassnett, Joel Faflak, Jo Devereux, Miranda Green-Barteet, and Alyssa MacLean. Kate Stanley and Mary Helen McMurran read chapters and provided much-needed emotional and intellectual support for this project. I had the amazing fortune to work with many brilliant graduate students at Western, whom I appreciate and admire as collaborators: Michael Sloane, Rasmus Simonsen, Andrew Weiss, Thomas Wormald, David Huebert, Jason Sandhar, and Josh Lambier. David provided astute comments on several chapters. The talented Jeff Swim gave the book a thorough full read. In Berlin, Jan Lensen and Jeremy Arnott helped me examine archives, translate German texts, and understand what happened there under the sign of extinction.

I have found kind support for ideas in this book in academic presentations hosted by Valli-Laurente Fraser-Celin, Ursula Heise, and Susie O'Brien. Gabriela Mastromonaco at the Toronto Zoo generously gave me a tour of the frozen zoo on site. Extended conversations with friends led to many aspects of this book, and I want to thank, in particular, Lynn Keller, Louis Cabri, Nicole Markotić, Dale Smith, Matthew Chrulew, Jonathan Skinner, Jenny Kerber, Matt Hooley, Angie Hume, Evelyn Reilly, David Goldstein, Catriona Sandilands, and Adam Dickinson. This book greatly benefited from comments and insights gleaned from John Riley, Deb Meert-Williston, Helen Fielding, Antonio Calcagno, Thomas Sorensen, Natalia Cecire, Leif Sorensen, Kate Marshall, Cary Wolfe, Doug Armato, Claire Colebrook, Richard Doyle, Gillian White, Ben Friedlander, Benjamin Hollander, Kristen Gallagher, Cheryl Lousley, Jennifer Baichwal, Nick de Pencier, Robert Folger, and everyone at CAPAS, Pierre-Louis Patoine, Olivier Brossard, and Frédéric Neyrat. From my Penn past, I want to thank Al Filreis, Paul Saint-Amour, Bob Perelman, Charles Bernstein, Heather Love, Jean-Michel (once more), everyone at the Kelly Writers House, Jessica Lowenthal, Julia Bloch, Andy Gaedtke, and Vance Bell. Benjy Kahan is always there as a friend, reader, and confidant. I also am very grateful for support while working on this book from extended family, and from friends who have become family: Leon Levin, Victor Levin, Ann Dychten-berg, Abigail Levin, Joe Kung, Jason Wolenik, Mark Litton, Nelson Sun, Peter Diaz, Evan Castel, and my St. Albans soccer peeps.

Fordham University Press has been a joy to work with. Richard Mor-rison is truly a special and generous editor. His care for this book has meant so much to me. I am thankful for all the attention and heedful-ness from the production team at Fordham. Thank you to Kathi Ander-son for copy editing. My two peer reviewers provided excellent insights and direction. I appreciate, in particular, the feedback from Susan McHugh and Ron Broglio.

Research for this book was supported by a grant from the Social Sci-ences and Humanities Research Council of Canada and grants for fac-ulty research from Western University. I am grateful for all the knowledge and expertise of the librarians at Western University and the many other libraries and archives I had the chance to visit during my research.

Some material from this book has appeared in previous publications: "Coral Cultures in the Anthropocene," *Cultural Review Studies* 25, no. 1 (2019), 85–102; "Life after Extinction," *Parrhesia* 27 (2017), 88–115; "Pho-tographing the Last Animal," *Antennae: The Journal of Nature in Visual Culture* 41 (2017), 102–22; "Sustainability after Extinction: On Last Ani-mals and Future Bison," in *Literature and Sustainability: Concept, Text, and Culture*, edited by Louise Squire, Adeline Johns-Putra, and John Parham (Manchester: Manchester University Press, 2017), 97–114; and "The Future of the Extinction Plot: Last Animals and Humans in Oc-tavia Butler's *Xenogenesis* Trilogy," *Humanimalia* 6, no. 2 (2015). I am deeply thankful to the editors of these publications: Matthew Chrulew and Rick De Vos, Arne De Boever, Giovanni Aloi, Louise Squire, and Istvan Csicsery-Ronay. I also want to thank the artists Kent Monk-man, Trevor Paglen, and Richard Vevers for permission to reproduce their work.

Some ideas for this book were worked out and composed with Derek Woods and published in our co-written book *Calamity Theory: Three Cri-tiques of Existential Risk* (University of Minnesota Press, 2021). It has a been a great pleasure to work with such a brilliant collaborator.

I feel fortunate to have the support and enthusiasm from my Califor-nia family with my brothers, Justin Schuster and Jordan Schuster, and my parents, Stewart and Bette Schuster, and my France family Francine and Roland Haddad, Igal Haddad, Laura Haddad, and Eva Haddad. This book is dedicated to three generations: my parents, my sons Reuven and Raphael, and my wonderful fleur Marina. To Marina, especially, *à la très-bonne, à la très-belle qui fait ma joie.*

# Notes

## Introduction

1. Population data from the International Union for the Conservation of Nature (IUCN) Red List, www.iucnredlist.org/, as of 2020.

2. M. Grooten and R. E. A. Almond, eds., "Living Planet Report—2018: Aiming Higher," *WWF* (Gland, Switzerland), https://c402277.ssl.cf1.rackcdn.com /publications/1187/files/original/LPR2018_Full_Report_Spreads.pdf.

3. Camille T. Dungy, *Trophic Cascade* (Middletown, CT: Wesleyan University Press, 2017), 33.

4. Francisco Sánchez-Bayo and Kris A. G. Wyckhuys, "Worldwide Decline of Entomofauna: A Review of Its Drivers," *Biological Conservation* 232 (April 2019), 8–27.

5. Carl Zimmer, "Birds Are Vanishing from North America," *New York Times*, September 19, 2019, www.nytimes.com/2019/09/19/science/bird-populations-america -canada.html.

6. Mukhisa Kituyi and Peter Thomson, "90% of Fish Stocks Are Used Up: Fisheries Subsidies Must Stop Emptying the Ocean," World Economic Forum, July 13, 2018, www.weforum.org/agenda/2018/07/fish-stocks-are-used-up-fisheries -subsidies-must-stop/.

7. Aelys M. Humphreys et al., "Global Dataset Shows Geography and Life Form Predict Modern Plant Extinction and Rediscovery," *Nature Ecology & Evolution* 3 (2019), 1043–47.

8. Samuel Taylor Coleridge, *The Complete Poems* (New York: Penguin, 1997), 186.

9. R. A. Philips et. al., "The Conservation Status and Priorities for Albatrosses and Large Petrels," *Biological Conservation* 201 (2016), 169–83.

10. Charles Darwin, *On the Origin of Species* (Oxford: Oxford University Press, 1996), 89.

11. Camilo Mora et al., "How Many Species Are There on Earth and in the Ocean?" *PLoS Biology* 9, no. 8 (2011), 1–8.

12. The estimate is from David M. Raup, *Extinction: Bad Genes or Bad Luck?* (New York: Norton, 1991), 4.

13. Ernst Mayer, *The Growth of Biological Thought: Diversity, Evolution, and Inheritance* (Cambridge, MA: Harvard University Press, 1982), 139.

14. David Sepkoski, *Rereading the Fossil Record: The Growth of Paleobiology as an Evolutionary Discipline* (Chicago: University of Chicago Press, 2012), 349.

15. David Sepkoski and Michael Ruse, ed., *The Paleobiological Revolution: Essays on the Growth of Modern Paleontology* (Chicago: University of Chicago Press, 2009).

16. Richard Leakey and Roger Lewin, *The Sixth Extinction: Patterns of Life and the Future of Humankind* (New York: Anchor Books, 1996); Elizabeth Kolbert, *The Sixth Extinction: An Unnatural History* (New York: Henry Holt, 2014).

17. Jurriaan M. De Vos et al., "Estimating the Normal Background Rate of Species Extinction," *Conservation Biology* 29, no. 2 (2014), 452–62; Stuart L. Pimm et al., "The Biodiversity of Species and Their Rates of Extinction, Distribution, and Protection," *Science* 344, no. 6187 (May 30, 2014); Rodolfo Dirzo et al., "Defaunation in the Anthropocene," *Science* 345, no. 6195 (July 25, 2014), 401–06; Mark J. Costello, "Biodiversity: The Known, Unknown, and Rates of Extinction," *Current Biology* 25 (May 4, 2015), 362–83.

18. "Nature's Dangerous Decline 'Unprecedented' Species Extinction Rates 'Accelerating,'" ipbes.net, www.ipbes.net/news/Media-Release-Global-Assessment. Accessed August 21, 2019.

19. Douglas Adams and Mark Carwardine, *Last Chance to See* (New York: Ballantine, 1990); Valmik Thapar, *The Last Tiger: Struggling for Survival* (Oxford: Oxford University Press, 2012); Trevor Paglen, *The Last Pictures* (Berkeley: University of California Press, 2012); Fred Bosworth, *Last of the Curlews* (Toronto: McLellan & Stewart, 2010); Don Pinnock and Colin Bell, *The Last Elephants* (Washington, DC: Smithsonian Books, 2019); Kate Brooks, dir., *The Last Animals* (Amazon, 2018); Dereck Joubert and Beverly Joubert, dirs. *The Last Lions* (Virgil Films, 2012); Steven Kazlowski, *The Last Polar Bear: Facing the Truth of a Warming World* (Seattle: Braided River, 2008); George B. Schaller, *The Last Panda* (Chicago: University of Chicago Press, 1994); Lawrence Anthony, with Graham Spence, *The Last Rhinos: My Battle to Save One of the World's Greatest Creatures* (New York: Thomas Dunne, 2012).

20. Jean M. O'Brien, *Firsting and Lasting: Writing Indians Out of Existence in New England* (Minneapolis: University of Minnesota Press, 2010).

21. Nicholas Mirzoeff, "It's Not the Anthropocene, It's the White Supremacy Scene," in *After Extinction*, ed. Richard Grusin (Minneapolis: University of Minnesota Press, 2018), 124.

22. Martin Crook and Damien Short, "Developmentalism and the Genocide-Ecocide Nexus," *Journal of Genocide Research* 23, no. 2 (2021), 162–88; Damien Short, *Redefining Genocide: Settler Colonialism, Social Death, and Ecocide* (London: Zed Books, 2016).

23. Winona LaDuke, *All Our Relations: Native Struggles for Land and Life* (Cambridge: South End Press, 1999), 1. See, also, David Harmon and Jonathan Loh, "Congruence between Species and Language Diversity," in *The Oxford Handbook of Endangered Languages*, ed. Kenneth L. Rehg and Lyle Campbell (Oxford: Oxford University Press, 2018), 659–82.

24. Rob Nixon, *Slow Violence and the Environmentalism of the Poor* (Cambridge, MA: Harvard University Press, 2011), 4.

25. Kyle Powys Whyte, "Indigenous Science (Fiction) for the Anthropocene: Ancestral Dystopias and Fantasies of Climate Change Crises," *Environment and Planning E: Nature and Space* 1, nos. 1–2 (2018), 226.

26. David Noble Cook, "Taino (Arawak) Indians," in *Encyclopedia of Genocide and Crimes Against Humanity*, vol 3., ed. Dinah L. Shelton (Detroit: Thomson Gale, 2015), 1017–19.

27. Ursula Heise, *Imagining Extinction: The Cultural Meanings of Endangered Species* (Chicago: University of Chicago Press, 2016), 5.

28. Deborah Bird Rose, *Wild Dog Dreaming: Love and Extinction* (Charlottesville: University of Virginia Press, 2011); Claire Colebrook, *Death of the PostHuman: Essays on Extinction*, vol. 1 (Open Humanities Press, 2015); Claire Colebrook, *Sex after Life: Essays on Extinction*, vol. 2 (Open Humanities Press, 2015); Mark V. Barrow, *Nature's Ghosts: Confronting Extinction from the Age of Jefferson to the Age of Ecology* (Chicago: University of Chicago Press, 2009); Heise, *Imagining Extinction*; Thom van Dooren, *Flight Ways: Life and Loss at the Edge of Extinction* (New York: Columbia University Press, 2014); Deborah Bird Rose, Thom van Dooren, and Matthew Chrulew, eds., *Extinction Studies: Stories of Time, Death, and Generations* (New York: Columbia University Press, 2017); Audra Mitchell, "Beyond Biodiversity and Species: Problematizing Extinction," *Theory, Culture and Society* 33 (2017), 23–42; Ashley Dawson, *Extinction: A Radical History* (New York: OR Books, 2016); *After Extinction*, ed. Richard Grusin (Minneapolis: University of Minnesota Press, 2018); Susan McHugh, *Love in a Time of Slaughters: Human-Animal Stories against Genocide and Extinction* (University Park: Penn State University Press, 2019); David Sepkoski, *Catastrophic Thinking: Extinction and the Value of Diversity from Darwin to the Anthropocene* (Chicago: University of Chicago Press, 2020); Sarah E. McFarland, *Ecocollapse Fiction and Cultures of Human Extinction* (London: Bloomsbury, 2021).

29. Charles Kingsley, *The Water Babies* (Peterborough: Broadview, 2008), 182.

30. Jacques Derrida, *The Work of Mourning*, ed. Pascale Anne-Brault and Michael Naas (Chicago: Chicago University Press, 2001), 65.

31. Kingsley's novel influenced Lewis Carroll to introduce the extinct dodo as a character in *Alice's Adventures in Wonderland* (1865).

32. Cited in Jeremy Gaskell, *Who Killed the Great Auk?* (Oxford: Oxford University Press, 2001), 71.

33. Thomas Moynihan discusses the distinction between apocalypse as a vision of theological judgment and extinction as a nonreligious natural event in *X-Risk: How Humanity Discovered Its Own Extinction* (Falmouth: Urbanomic, 2020).

34. Arthur Lovejoy, *The Great Chain of Being: A Study of the History of an Idea* (Cambridge, MA: Harvard University Press, 1936), 52.

35. Aristotle, *De Anima*, tr. Hugh Lawson-Tancred (New York: Penguin, 1986), 218.

36. Aristotle, *Progression of Animals*, tr. E. S. Forster (Cambridge, MA: Harvard University Press, 1961), 487.

37. John G. T. Anderson, *Deep Things Out of Darkness: A History of Natural History* (Berkeley: University of California Press, 2013), 33.

38. Robert Silverberg, *The Auk, The Dodo, and the Oryx: Vanished and Vanishing Creatures* (New York: Thomas Crowell, 1967), 7

39. Cited in Mary Louise Pratt, *Imperial Eyes: Travel Writing and Transculturation* (London: Routledge, 1992), 30.

40. Jacques Roger, *Buffon: A Life in Natural History*, tr. Sarah Lucille Bonnefoi (Ithaca: Cornell University Press, 1997), 307.

41. A proposal for the perishability of some species appears already in the writings of Lucretius, but he does not develop a theory of extinction as an ongoing biological reality. Lucretius claimed that in the early period of the Earth a number of "monstrous" animals roamed and "Many species must have died out altogether and failed to reproduce their kind." Lucretius, *The Nature of the Universe*, tr. R. E Latham (Baltimore: Penguin, 1962), 197. In Buffon's own time, the French encyclopedist Denis Diderot considered that it might be the case that humans would go extinct for a time, but after "at the end of several hundreds of million years and of I-don't-know-whats, the biped animal who carries the name man" would arise again. Cited in Rebecca Stotts, *Darwin's Ghosts: The Secret History of Evolution* (New York: Spiegel & Grau, 2012), 148.

42. Martin J. S. Rudwick, ed., *Georges Cuvier, Fossil Bones, and Geological Catastrophes: New Translations and Interpretations of the Primary Texts* (Chicago: University of Chicago Press, 1997), 22.

43. Georges Cuvier, *Essay on the Theory of the Earth*, tr. Robert Kerr (Cambridge: Cambridge University Press, 2009 [1815 reprint]), 7.

44. In 1816, Cuvier sought to examine the body of Sara Bartmaan, a Black Khoikhoi woman from southwestern Africa who had been exhibited as a curiosity in Europe. Cuvier dissected Bartmaan's body, focusing on her sexual organs, and kept her skeletal remains as a museum piece, declaring her as evidence of a lower racial typology with anatomical traits similar to apes. See Clifton C. Crais and Pamela Scully, *Sara Baartman and the Hottentot Venus: A Ghost Story and Biography* (Princeton: Princeton University Press, 2009).

45. Michel Foucault, *The Order of Things: An Archaeology of the Human Sciences* (New York: Routledge, 2002), 180.

46. Honoré de Balzac, *The Wild Ass's Skin*, tr. Helen Constantine (Oxford: Oxford University Press, 2012), 19.

47. Charles Lyell, *Principles of Geology*, ed. James Secord (London: Penguin, 1997 [reprint of first edition 1830–33]), 277.

48. Darwin, *On the Origin of Species*, 257.

49. Cited in Richard Holmes, *The Age of Wonder* (New York: Vintage, 2008), 209.

50. William Whewell, *Astronomy and General Physics Considered with Reference to Natural Theology* (London: William Pickering, 1839), 203–4. Richard Holmes summarizes Herschel's conclusion as stating that "Our solar system, our planet, and hence our whole civilization would have an ultimate and unavoidable end." Holmes, *Age of Wonder*, 209.

51. Balzac, *Wild Ass's Skin*, 19–20. Translation modified.

52. See David N. Stamos, *The Species Problem: Biological Species, Ontology, and the Metaphysics of Biology* (Lanham, MD: Lexington Books, 2004); Robert J. Richards, *The Species Problem: A Philosophical Analysis* (Cambridge: Cambridge University Press, 2010).

53. Darwin, *On the Origin of Species*, 38.

54. Richard Dawkins, *The Selfish Gene* (Oxford: Oxford University Press, 2006), vii.

55. Donna Haraway, *Simians, Cyborgs, and Women: The Reinvention of Nature* (New York: Routledge, 1991), 200.

56. Stephanie LeMenager mentions once asking her students "if they 'feel like a species'" and was surprised that several stated "they feel like a species only insofar as they imagine themselves at the edge of extinction." LeMenager, "The Humanities after the Anthropocene," in *The Routledge Companion to the Environmental Humanities*, ed. Ursula K. Heise, Jon Christensen, and Michelle Niemann (New York: Routledge, 2017), 478.

57. Eileen Crist, *Abundant Earth: Toward an Ecological Civilization* (Chicago: University of Chicago Press, 2019), 17.

58. Consider, however, that without extinction, biodiversity would proliferate exponentially, checked only by individual deaths. This would make the concept of biodiversity rather useless. Extinction and biodiversity require each other to make sense of the proliferation and fragility of the species form. Ursula Heise discusses the ambiguities of the notion of biodiversity in connection with extinction in "Lost Dogs, Last Birds, and Listed Species: Cultures of Extinction," *Configurations* 18, nos. 1–2 (Winter 2010), 49–72.

59. Ray Brassier, *Nihil Unbound: Enlightenment and Extinction* (New York: Palgrave Macmillan, 2007), 234.

60. Brassier, *Nihil Unbound*, 229.

61. For a further analysis of the conceptual limitations of vitalism and reductionist materialism, see Joshua Schuster, "Life after Extinction," *Parrhesia* 27 (2017), 88–115.

62. See, for example, Frederick Buell, *From Apocalypse to Way of Life: Environmental Crisis in the American Century* (New York: Routledge, 2003); Alan Weisman, *The World without Us* (New York: Thomas Dunne, 2007); Eva Horn, *The Future as Catastrophe: Imagining Disaster in the Modern Age* (New York: Columbia University Press, 2018).

63. See Ronald L. Sandler, *The Ethics of Species: An Introduction* (Cambridge: Cambridge University Press, 2012), 29.

64. See Ashlee Cunsolo and Karen Landman, ed., *Mourning Nature: Hope at the Heart of Ecological Loss and Grief* (Kingston: McGill-Queen's University Press, 2017); Owain Jones, Kate Rigby, and Linda Williams, "Everyday Ecocide, Toxic Dwelling, and the Inability to Mourn: A Response to Geographies of Extinction," *Environmental Humanities* 12, no. 1 (May 2020), 388–405.

65. Theodora Kroeber, *Ishi in Two Worlds: A Biography of the Last Wild Indian in North America* (Berkeley: University of California Press, 1961).

66. Hannah Arendt, *Eichmann in Jerusalem: A Report on the Banality of Evil* (New York: Penguin, 2006), 279. See, also, Kelly Oliver, *Earth and World: Philosophy after the Apollo Missions* (New York: Columbia University Press, 2015). Arendt's reflections are influenced by Kant's statement in "Perpetual Peace" that everyone has the "right to the earth's surface which the human race shares in common." Kant adds: "Since the earth is a globe, [people] cannot disperse over an infinite area, but must necessarily tolerate one another's company. And no-one originally has any greater right than any one else to occupy any particular portion of the earth." Immanuel Kant, *Political Writings*, 2nd edition, tr. H. B. Nisbet (Cambridge: Cambridge University Press, 1991), 106.

67. Hannah Arendt, *The Human Condition* (Chicago: University of Chicago Press, 1998), 234.

68. W. S. Merwin, *The Lice* (New York: Atheneum, 1967), 68.

69. Heise, *Imagining Extinction*, 47.

70. Hank Lazer, "For a Coming Extinction: A Reading of W. S. Merwin's *The Lice*," *ELH* 49, no. 1 (1982), 262.

71. Paul Sheehan, "Myth, Absence, Haunting: Toward a Zoopoetics of Extinction," in *What is Zoopoetics? Texts, Bodies, Entanglement*, ed. Kári Driscoll and Eva Hoffman (New York: Palgrave, 2018), 178–79.

72. Eleni Sikelianos, *Make Yourself Happy* (Minneapolis: Coffee House Books, 2017), 69.

73. Gregory Nagy, "Ancient Greek Elegy," in *The Oxford Handbook of the Elegy*, ed. Karen Weisman (Oxford: Oxford University Press, 2010), 24.

74. Linda Hogan, *Dwellings: A Spiritual History of the Living World* (New York: Simon and Schuster, 1996), 115.

75. Rose, *Wild Dog Dreaming*, 42.

## 1 / Photographing the Last Animal

1. Frederick Jackson Turner, *The Frontier in American History* (Mineola, NY: Dover, 2010), 38.

2. David A. Dary, *The Buffalo Book: The Full Saga of the American Animal* (Chicago: Sage Books, 1974), 122.

3. The American bison (*bison bison*) is descended from the *bison* genus but has also long been called a buffalo, leading to some confusion with the genus *bubalus*, which includes the water buffalo. Both are in the *bovidae* family. Bison and buffalo still are commonly used as interchangeable names today, though taxonomists prefer bison.

4. Joel Allen, *History of the American Bison, Bison Americanus* (Washington, DC: Government Printing Office, 1877), 556.

5. John Berger, *About Looking* (New York: Vintage, 1991), 26.

6. Andrew C. Isenberg, *The Destruction of the Bison: An Environmental History* (Cambridge: Cambridge University Press, 2000), 25; see, also, Dan Flores, "Bison Ecology and Bison Diplomacy: The Southern Plains from 1800 to 1850," *Journal of American History* 78 (September 1991), 465–85; Dale F. Lott, *American Bison: A Natural History* (Berkeley: University of California Press, 2002).

7. Antoine Traisnel, *Capture: American Pursuits and the Making of a New Animal Condition* (Minneapolis: University of Minnesota Press, 2020), 4.

8. In addition to Roe and Isenberg, see William T. Hornaday, "The Extermination of the American Bison," *Report of the United States National Museum for 1887* (Washington, DC: Government Printing Office, 1889); Martin S. Garretson, *A Short History of the American Bison* (Freeport, NY: Books for Libraries Press, 1934); J. Albert Rorabacher, *The American Buffalo in Transition: A Historical and Economic Survey of the Bison in America* (St. Cloud, MN: North Star Press, 1970); Charles M. Robinson III, *The Buffalo Hunters* (Abilene: State House Press, 1995).

9. Isenberg makes the important point that there is considerable fluctuation within an animal population even when it is considered relatively stable: "The nineteenth-century bison population was not static but constantly in flux." Isenberg, *Destruction of the Bison*, 26. Isenberg lists a wide range of changing ecological factors, including variable weather, drought, population of wolf predators, disease,

competition for food by wild horses, and mortality and birth rates for bison. These variables all play a role in exacerbating the effects of human hunting practices on the bison population.

10. Frank Gilbert Roe, *The North American Buffalo: A Critical Study of the Species in its Wild State,* 2nd edition (Toronto: University of Toronto Press, 1970), 437.

11. Roe, *North American Buffalo,* 441.

12. Isenberg, *Destruction of the Bison,* 137.

13. For a detailed study of the camera technology used in nineteenth-century photography of the American West, see Martha Sandweiss, *Print the Legend: Photography and the American West* (New Haven: Yale University Press, 2002).

14. Darwin, *On the Origin of Species,* 260.

15. Luther Standing Bear, *My People the Sioux* (Boston: Houghton Mifflin, 1928), 30.

16. Charles A. Eastman, *The Indian To-day: The Past and Future of the First American* (Garden City, NY: Doubleday, Page & Co., 1915), 32.

17. Old Lady Horse (Spear Woman), "The End of the World: The Buffalo Go," in *Legends of Our Times: Native Cowboy Life,* ed. Morgan Baillargeon and Leslie Tepper (Vancouver: UBC Press, 1998), 129. Many plains Indigenous communities used the bison skin, shaped into teepees and clothing, as an artistic medium on which to create pictographs and tell stories, including stories of hunting the bison. Drawings made in the nineteenth century in ledger books by Indigenous peoples also included depictions of bison hunts. See Evan M. Mauer, ed., *Visions of the People: A Pictorial History of Plains Indian Life* (Minneapolis: Minneapolis Institute of Arts, 1992); Janet Catherine Berlo, *Spirit Beings and Sun Dancers: Black Hawk's Vision of the Lakota World* (New York: George Braziller, 2000).

18. Cited in Collin G. Calloway, ed., *Our Hearts Fell to the Ground: Plains Indians Views of How the West Was Lost* (New York: Bedford Books, 1996), 123.

19. Tasha Hubbard, "Buffalo Genocide in Nineteenth-Century North America: 'Kill, Skin, and Sell,'" in *Colonial Genocide in Indigenous North America,* ed. Andrew Woolford, Jeff Benvenuto, and Alexander Laban Hinton (Durham: Duke University Press, 2014), 301.

20. Dary, *Buffalo Book,* 127.

21. "Letter from the Secretary of the Interior, Communicating . . . information in relation to the late massacre of United States troops by Indians at or near Fort Phil. Kearney, in Dakota Territory," Senate Executive Document 13, 40th Congress, First Session. Cited in Isenberg, *Destruction of the Bison,* 124.

22. Hornaday, "The Extermination of the American Bison," 484.

23. Ibid., 521. The question mark is Hornaday's.

24. Larry Len Peterson, *L. A. Huffman: Photographer of the American West,* 2nd edition (Missoula: Mountain Press, 2013), xi.

25. Thanks to Jeff Swim for this suggestion.

26. Carol J. Adams, *The Sexual Politics of Meat: A Feminist-Vegetarian Critical Theory* (New York: Bloomsbury, 2015).

27. Giorgio Agamben, *Homo Sacer: Sovereign Power and Bare Life,* tr. Daniel Heller-Roazen (Stanford: Stanford University Press, 1998).

28. Nicole Shukin, *Animal Capital: Rendering Life in Biopolitical Times* (Minneapolis: University of Minnesota Press, 2009).

29. LeRoy Barnett, "Buffalo Bones in Detroit," *Detroit in Perspective* 2, no. 2 (1975), 91.

30. Matthew Brower, *Developing Animals: Wildlife and Early American Photography* (Minneapolis: University of Minnesota Press, 2011), 40.

31. Alfred W. Crosby, *Ecological Imperialism: The Biological Expansion of Europe, 900–1900*, 2nd edition (Cambridge: Cambridge University Press, 2004).

32. For further discussion of the visual legacy of "last bison" imagery, see Danielle Tascherau Mamers, "'Last of the Buffalo': Bison Extermination, Early Conservation, and Visual Records of Settler Colonization in the North American West," *Settler Colonial Studies* 10, no. 1 (2020), 126–47. For a broader discussion of the role of photography in colonialism, see Daniel Foliard, *The Violence of Colonial Photography* (Manchester: Manchester University Press, 2023).

33. Susan Sontag, *Regarding the Pain of Others* (New York: Farrar, Straus and Giroux, 2003), 24

34. André Bazin, *What Is Cinema?* vol. 1, tr. Hugh Gray (Berkeley: University of California Press, 2005), 10.

35. Joanna Zylinska discusses how Bazin's "embalming time" is an early insight toward how the impassive mechanical camera eye "sees" from the perspective of "deep history" and species extinction. Joanna Zylinska, "Photography after Extinction," in *After Extinction*, ed. Richard Grusin, 51–70.

36. Susan Sontag, *On Photography* (New York: Anchor Books: 1977), 67.

37. Roland Barthes, *Camera Lucida: Reflections on Photography*, tr. Richard Howard (New York: Noonday Press, 1981), 92.

38. Peggy Phelan, "Atrocity and Action: The Performative Force of the Abu Ghraib Photographs," in *Picturing Atrocity: Photography in Crisis*, ed. Geoffrey Batchen, Mick Gidley, Nancy K. Miller, and Jay Prosser (London: Reaktion Books, 2012), 55. An extended discussion of photography and trauma can be found in Ulrich Baer, *Spectral Evidence: The Photography of Trauma* (Cambridge, MA: MIT Press, 2002). For further reflection on Freud's use of the photographic metaphor and the extension of Freudian terms throughout the photographic field, see Shawn Michelle Smith and Sharon Sliwinski, ed., *Photography and the Optical Unconscious* (Durham: Duke University Press, 2017).

39. Barthes, *Camera Lucida*, 96.

40. Douglas Adams and Mark Carwardine, *Last Chance to See* (New York: Ballantine, 1990); Errol Fuller, *Lost Animal: Extinction and the Photographic Record* (London: Bloomsbury, 2013); Joel Sartore, *Rare: Portraits of America's Endangered Species* (Washington, DC: National Geographic, 2010).

41. Sontag, *On Photography*, 88.

## 2 / Indigeneity and Anthropology in Last Worlds

1. Theodora Kroeber, *Ishi in Two Worlds: A Biography of the Last Wild Indian in North America* (Berkeley: University of California Press, 1961).

2. Gerald Vizenor and A. Robert Lee, *Postindian Conversations* (Lincoln: University of Nebraska, 1999), 88.

3. Gerald Vizenor, *Manifest Manners: Narratives on Postindian Survivance* (Lincoln: University of Nebraska, 1999), 133.

4. Orin Starn, *Ishi's Brain: In Search of America's "Last" Wild Indian* (New York: Norton, 2004), 27.

5. Benjamin Madley, *An American Genocide: The United States and the California Indian Catastrophe* (New Haven: Yale University Press), 173.

6. Charles Darwin, *The Descent of Man* (Amherst: Prometheus Books, 1998 [reprint second edition 1874]), 189.

7. Ward Churchill and Winona LaDuke, *Struggle for the Land: Native North American Resistance to Genocide, Ecocide, and Colonization* (San Francisco: City Lights, 2002), 54.

8. Daryl Baldwin, Margaret Noodin, and Bernard C. Perley, "Surviving the Sixth Extinction," in *After Extinction*, ed. Richard Grusin (Minneapolis: University of Minnesota Press, 2018), 213.

9. Kroeber, *Ishi in Two Worlds*, 229.

10. Alfred Kroeber, "The Only Man in America Who Knows No Christmas," in *Ishi the Last Yahi: A Documentary History*, ed. Robert F. Heizer and Theodora Kroeber (Berkeley: University of California Press, 1979), 112.

11. Jean M. O'Brien, *Firsting and Lasting: Writing Indians Out of Existence in New England* (Minneapolis: University of Minnesota Press, 2010), 105. For a further discussion of the colonialist claim of firstness, see Lauren Beck, ed., *Firsting in the Early-Modern Atlantic World* (New York: Routledge, 2020).

12. Daniel Heath Justice, *Why Indigenous Literatures Matter* (Waterloo: Wilfrid Laurier Press, 2018), 4.

13. Kroeber, *Ishi in Two Worlds*, 98.

14. Jacques Derrida, *The Post Card: From Socrates to Freud and Beyond*, tr. Alan Bass (Chicago: University of Chicago Press, 1979).

15. Theodora Kroeber, "Acknowledgments," *Ishi the Last Yahi: A Documentary History*, n.p.

16. Alfred Kroeber, "Eighteen Professions," *American Anthropologist* 17, no. 3 (1915), 284. Eric Wolf critiques Kroeber's elevation of the group as the foundational anthropological unit of study, remarking that in Kroeber's writings, "there are, in fact, no people." Eric Wolf, "Alfred L. Kroeber," in *Totems and Teachers: Perspectives on the History of Anthropology*, ed. Sydel Silverman (New York: Columbia University Press, 1981), 57.

17. Shari Huhndorf, *Going Native: Indians in the American Cultural Imagination* (Ithaca: Cornell University Press, 2001), 14.

18. See, for example, Orin Starn, "Ishi's Spanish Words," in *Ishi in Three Centuries*, ed. Karl Kroeber and Clifton Kroeber (Lincoln: University of Nebraska Press, 2008), 201–7.

19. Linda Tuhiwai Smith, *Decolonizing Methodologies: Research and Indigenous Peoples* (London: Zed Books, 1999), 74.

20. *Ishi the Last Yahi: A Documentary History*, 96.

21. Richard Bernheimer, *Wild Men in the Middle Ages: A Study in Art, Sentiment, and Demonology* (Cambridge, MA: Harvard University Press, 1952).

22. Patrick Brantlinger, *Dark Vanishings: Discourse on the Extinction of Primitive Races, 1800–1930* (Ithaca: Cornell University Press, 2003), 4.

23. David Wallace Adams, *Education or Extinction: American Indians and the Boarding School Experience, 1875–1928* (Lawrence: University Press of Kansas, 1995), 16.

24. Douglas Sackman, *Wild Men: Ishi and Kroeber in the Wilderness of Modern America* (Oxford: Oxford University Press, 2010), 309.

25. Alfred Kroeber, "Ishi the Last Aborigine: The Effects of Civilization on a Genuine Survivor of Stone Age Barbarism," *The World's Work* (July 1912), 304–8.

26. Ira Jacknis, "'The Last Wild Indian in North America': Changing Museum Representations of Ishi," in *Museums and Difference*, ed. Daniel J. Sherman (Bloomington: Indiana University Press, 2008), 69.

27. For more detail on these photography sessions, see Richard L. Burrill, *Ishi's Untold Story in His First World* (Red Bluff: Paragon Publishing, 2011), 121.

28. The image of Ishi partially clothed draws on a long history of associating indigeneity with unclothedness. Philippa Levine remarks that "a lack of clothing among colonized individuals has connoted primitiveness and savagery since at least the seventeenth century." She adds, "undress, at least in the context of the long nineteenth century, can be adequately theorized only through the lens of colonialism." Philippa Levine, "States of Undress: Nakedness and the Colonial Imagination," *Victorian Studies* 50, no. 2 (2008), 189, 191.

29. Brian Hochman, *Savage Preservation: The Ethnographic Origins of Modern Media Technology* (Minneapolis: University of Minnesota Press, 2014), xiii. For an extended study of one collection of "salvage" ethnographic photographs that attends to the aesthetic, technological, and ethical ramifications of such photos, see Shamoon Zamir, "The Gift of the Face: Portraiture and Time," in Edward S. Curtis, *The North American Indian* (Chapel Hill: University of North Carolina Press, 2014).

30. Madley, *An American Genocide*, 3.

31. Kroeber, *Ishi in Two Worlds*, 72.

32. Ibid., 111.

33. Pauline Wakeham, *Taxidermic Signs: Reconstructing Aboriginality* (Minneapolis: University of Minnesota Press, 2008), 21.

34. Kenn Harper, *Minik: The New York Eskimo: An Artic Explorer, a Museum, and the Betrayal of the Inuit People*, revised edition (South Royalton, VT: Steerforth Press, 2017).

35. Jace Weaver, "When the Demons Come: (Retro)Spectacle among the Savages," in *Ishi in Three Centuries*, 45. Weaver mentions that William Temple Hornaday, who had led efforts to preserve bison, as the director of the Bronx Zoological Gardens approved of exhibiting the Congolese man Ota Benga in a cage. Theodora Kroeber writes of both Ota Benga and the Inuit Greenlanders who suffered at the hands of anthropologists and museum directors. Theodora Kroeber insists Alfred Kroeber was at pains to avoid exposing Ishi to similar mistreatments.

36. Kroeber, "Ishi the Last Aborigine," in *Ishi the Last Yahi: A Documentary History*, 122.

37. Kroeber, "The Yana and Yahi," in *Ishi the Last Yahi: A Documentary History*, 166.

38. Vizenor, *Manifest Manners*, 128.

39. Kroeber, *Ishi in Two Worlds*, 137.

40. Ira Jacknis, "'The Last Wild Indian in North America': Changing Museum Representations of Ishi," 70.

41. See Scott Lauria Morgensen, *Spaces between Us: Queer Settler Colonialism and Indigenous Decolonization* (Minneapolis: University of Minnesota Press, 2011); Qwo-Li Driskill, Chris Finley, Brian Joseph Gilley, Scott Lauria Morgensen, ed., *Queer Indigenous Studies: Critical Interventions in Theory, Politics, and Literature* (Tucson: University of Arizona Press, 2011).

42. Kroeber, *Ishi in Two Worlds*, 234.

43. Starn, *Ishi's Brain*, ch. 9.

44. Sigmund Freud, *The Standard Edition of the Complete Psychological Works of Sigmund Freud*, vol. XIV *(1914–1916)*, tr. James Strachey (London: Hogarth Press, 1957), 168.

45. Frank J. Sulloway, *Freud: Biologist of the Mind* (Cambridge, MA: Harvard University Press, 1992).

46. Freud, *The Standard Edition, vol. XIV (1914–1916)*, 286.

47. Sigmund Freud, *The Standard Edition of the Complete Psychological Works of Sigmund Freud*, vol. XIII *(1913–1914)*, tr. James Strachey (London: Vintage, 2001), 21.

48. Christian Kerslake, "Introduction," in *Origins and Ends of the Mind: Philosophical Essays on Psychoanalysis*, ed. Christian Kerslake and Ray Brassier (Leuven: Leuven University Press, 2007), 2.

49. John C. Burnham, "Anthropologist A. L. Kroeber's Career as a Psychoanalyst: New Evidence and Lessons from a Significant Case History," *American Imago* 69, no. 1 (Spring 2012), 17.

50. Alfred Kroeber, *Anthropology* (New York: Harcourt, Brace, and Company 1923).

51. Freud, *The Standard Edition*, vol. XIII, 18.

52. Alfred Kroeber, *The Nature of Culture* (Chicago: University of Chicago Press, 1952), 304.

53. A footnote in *Group Psychology* clarifies that Freud is confusing the "just-so story" accusation with another review of the book where that phrase appears written by the British anthropologist R. R. Marrett. In the first edition of *Group Psychology*, Freud mentions the name "Kroeger" in the text, a mistaken reference to Kroeber. He removed the name altogether in subsequent editions. Sigmund Freud, *Group Psychology and the Analysis of the Ego*, tr. James Strachey (New York: Norton, 1989), 69.

54. Jacques Derrida, *Archive Fever: A Freudian Impression*, tr. Eric Prenowitz (Chicago: University of Chicago Press, 1996), 85.

55. Sigmund Freud, *Beyond the Pleasure Principle*, ed. Todd Dufresne, tr. Gregory C. Richter (Peterborough: Broadview, 2011), 75.

56. Paul Ricoeur, "Open Questions: On Negation, Pleasure, Reality," in "Appendix B," in *Beyond the Pleasure Principle*, ed. Todd Dufresne, 214.

57. Brassier, *Nihil Unbound*, 235.

58. Sigmund Freud, *The Standard Edition of the Complete Psychological Works of Sigmund Freud*, vol. XXI *(1927–31)*, tr. James Strachey (London: Hogarth Press, 1961), 145.

59. Freud, *Beyond the Pleasure Principle*, 58.

60. Compare his more apocalyptic reflections in the short essay "On Transience" (1916): "Limitation in the possibility of enjoyment raises the value of enjoyment. . . . The beauty of the human form and face vanish for ever in the course of our own lives, but their evanescence only lends them a fresh charm. . . . A time may indeed come when the pictures and statues which we admire to-day will crumble to dust, or a race of men may follow us who no longer understand the works of our poets and thinkers, or a geological epoch may even arrive when all animate life upon the earth ceases; but since the value of all this beauty and perfection is determined only by its significance for our own emotional lives, it has no need to survive us and is there independent of absolute duration." Freud, *The Standard Edition*, vol. XIV, 305–6.

61. The phrase is from an unpublished manuscript, now available as Sigmund Freud, *A Phylogenetic Fantasy: Overview of the Transference Neuroses*, ed. Ilse Grubrich-Simitis, tr. Alex Hoffer and Peter T. Hoffer (Cambridge, MA: Harvard University Press, 1987).

62. Freud, *The Standard Edition*, vol. XIII, 1.

63. Ranjana Khanna, *Dark Continents: Psychoanalysis and Colonialism* (Durham: Duke University Press, 2003), 6. See, also, Celia Brickman, *Race in Psychoanalysis: Aboriginal Populations in the Mind* (New York: Routledge, 2018).

64. Julia V. Emberley, *Defamiliarizing the Aboriginal: Cultural Practices and Decolonization in Canada* (Toronto: University of Toronto Press, 2007), 108.

65. Stef Craps argues that the field of trauma studies that developed in the 1990s also has continued the Eurocentric tendencies of Freud and requires reorientation toward postcolonial and anti-colonial understandings of different regional histories of trauma and its psychic manifestations. Stef Craps, *Postcolonial Witnessing: Trauma Out of Bounds* (New York: Palgrave, 2013).

66. Leilani Salvo Crane, "Invisible: A Mixt Asian Woman's Efforts to See and Be Seen in Psychoanalysis," *Studies in Gender and Sexuality* 21, no. 2 (2020), 128.

67. Kroeber, *Handbook of the Indians of California*, 464. The Muwekma Ohlone Tribe of the San Francisco Bay Area, which still has not obtained federal status, continues to petition for recognition. Sabrina Imbler, "New DNA Analysis Supports an Unrecognized Tribe's Ancient Roots in California," *New York Times* (April 12, 2022), https://www.nytimes.com/2022/04/12/science/muwekma-ohlone-tribe -california-dna.html.

68. Nancy Scheper-Hughes, "Ishi's Brian, Ishi's Ashes: Reflections on Anthropology and Genocide," in *Ishi in Three Centuries*, 100.

69. Kroeber, *The Nature of Culture*, 314.

70. Kroeber, *Anthropology*, 12.

71. Roxanne Dunbar-Ortiz, *An Indigenous Peoples' History of the United States* (Boston: Beacon Press, 2014), iii.

72. James Clifford, *Returns: Becoming Indigenous in the Twenty-First Century* (Cambridge, MA: Harvard University, 2013), 91.

73. Karen Biestman, "Ishi and the University," in *Ishi in Three Centuries*, 153.

## 3 / Literary Extinctions and the Existentiality of Reading

1. H. G. Wells, *The Discovery of the Future* (New York: B. W. Huebsch, 1913), 54–56.

2. William Morris, in the subtitle for his *News from Nowhere*, called the genre a "utopian romance"—Wells's work, then, appears dialectically as the dystopian romance, although the term "dystopia" was coined several decades later. William Morris, *News from Nowhere and Other Writings* (New York: Penguin, 1993).

3. Mike Davis, *Late Victorian Holocausts: El Niño Famines and the Making of the Third World* (London: Verso Books, 2001).

4. Gillian Beer, *Darwin's Plots: Evolutionary Narrative in Darwin, George Eliot and Nineteenth-Century Fiction*, 2nd edition (Cambridge: Cambridge University Press, 2000).

5. Daniel Dennett, *Darwin's Dangerous Idea: Evolution and the Meanings of Life* (New York: Touchstone, 1995).

6. Charles Darwin, *Journal of Researches into the Natural History and Geology of the Countries Visited During the Voyage of H. M. S. Beagle Round the World*, 2nd edition (London: John Murray, 1845), 174.

7. Charles Darwin, *The Descent of Man; and Selection in Relation to Sex*, 2nd edition (London: John Murray, 1874), 6.

8. Darwin, *Descent of Man*, 190.

9. Sylvia Wynter, "Unsettling the Coloniality of Being/Power/Truth/Freedom," *CR: The New Centennial Review* 3, no. 3 (2003), 267.

10. Claire Colebrook, *Death of the PostHuman: Essays on Extinction*, vol. 1 (Ann Arbor: Open Humanities Press, 2014), 39.

11. Keith Williams, *H. G. Wells, Modernity, and the Movies* (Liverpool: Liverpool University Press, 2007), 3.

12. David Wittenberg, *Time Travel: The Popular Philosophy of Narrative* (New York: Fordham University Press, 2013), 87–88.

13. H. G. Wells, *The Time Machine* (New York: Penguin Books, 2005 [1895]), 83.

14. H. G. Wells, *Early Scientific Writings and Essays*, ed. Robert M. Philmus and David Y. Hughes (Berkeley: University of California Press, 1975), 169.

15. H. G. Wells, "The Extinction of Man: Some Speculative Suggestions," *Pall Mall Gazette* 59 (September 25, 1894), 3.

16. Max Nordau, *Degeneration* (London: William Heinmann, 1895), 2. Nordau thought the figure of the dying human emblematic not of planetary ecological ruin—he believed nature broadly to be still a paragon of health—but of a certain view of European history in decline.

17. Christina Alt, "Extinction, Extermination, and the Ecological Optimism of H. G. Wells," in *Green Planets: Ecology and Science Fiction*, ed. Gerry Canavan and Kim Stanley Robinson (Middletown, CT: Wesleyan University Press, 2014), 37.

18. Robert Crossley notes that the materiality of natural history museums recurs frequently in science fiction narratives: "artefacts from the museums, libraries, and archives of science fiction provides variants of a scene that gets reconstituted with astonishing frequency in science-fiction narratives. The spectacle of an observer examining an artefact and using it as a window on to nature, culture, and history permits that convergence of anthropological, prophetic, and elegiac tonalities that science fiction handles more powerfully than any other modern literary form." Robert Crossley, "In the Palace of Green Porcelain: Artefacts from the Museums of Science Fiction," *Science Fiction Essays and Studies* 43 (1990), 77–78.

19. "In 1890, R. P. Cameron, writing to support the reform of museums in accordance with evolutionary principles, complained that the popular idea of a museum was still that of 'a sort of charnel house for dead animals, skeletons and skulls; that it was a dungeon-like place, dark, dusty and dreary.'" Tony Bennett, *Pasts Beyond Memory: Evolution, Museums, Colonialism* (London: Routledge, 2004), 13.

20. Anonymous, *A Guide to the Exhibition Galleries of the Department of Geology and Paleontology in the British Museum (Natural History)* (London: Order of the Trustees, 1890), 68.

21. Wittenberg, *Time Travel*, 2.

22. Wells, *Time Machine*, 41–42.

23. Brian Richardson, *Unnatural Narrative: Theory, History, and Practice* (Columbus: Ohio State University Press, 2015), 3.

24. Brian Richardson, *Unnatural Voices: Extreme Narration in Modern and Contemporary Fiction* (Columbus: Ohio State University Press, 2006), 1. See, also, David Herman, *Narratology Beyond the Human: Storytelling and Animal Life* (Oxford: Oxford University Press, 2018).

25. Richardson, *Unnatural Voices*, 87.

26. Frank Kermode, in *The Sense of an Ending: Studies in the Theory of Fiction* (Oxford: Oxford University Press, 1967), provided the first formalist and narratological study of apocalyptic stories, noting they are common across literary history and correspond to the human need for beginnings, middles, and ends. Kermode reads ends, apocalyptic or not, as necessary and normative to literary form. Lecia Rosenthal, in her study of catastrophic tropes in modernism, finds an unresolvable tension between literary norms of endings and closure and the destructivity of apocalyptic events: "With its connotations of explosive negativity, unredeemed violence, and unprecedented powers of destruction, catastrophe carries the burden of signifying a history that exceeds, and for that reason continues to demand, the reassuring claims of ethical closure and narrative coherence." Lecia Rosenthal, *Mourning Modernism: Literature, Catastrophe, and the Politics of Consolation* (New York: Fordham University Press, 2011), 22.

27. Marco Caracciolo, *Strange Narrators in Contemporary Fiction: Exploration in Readers' Engagement with Characters* (Lincoln: University of Nebraska, 2016).

28. Marie-Laure Ryan, "Possible Worlds to Storyworlds: On the Worldness of Narrative Representations," in *Possible Worlds Theory and Contemporary Narratology*, ed. Alice Bell and Marie-Laure Ryan (Lincoln: University of Nebraska Press, 2019), 67.

29. Jan Alber, *Unnatural Narrative: Impossible Worlds in Fiction and Drama* (Lincoln: University of Nebraska Press, 2016), 7.

30. Martin McQuillan demonstrates the paradox aptly: "What if there were no more stories? What if stories, contrary to narrative theory's fundamental article of faith (that narrative, like love, is all around us), were coming, or had already come, to an end? How would we know? What would we do? How would we identify or constitute the 'we' of these questions? If stories (or Story, the idea of Story) had already come to an end it would not be that simple a task to declare it so." Martin McQuillan, "Introduction: Aporias of Writing: Narrative and Subjectivity," in *The Narrative Reader*, ed. Martin McQuillan (New York: Routledge, 2000), 1.

31. E. M. Forster, *Aspects of the Novel*, (New York: Harcourt, Brace & World, 1954 [1927]), 72.

32. James Phelan claims that a character must be linked with progression of plot. In Wells's novella, the Time Traveller does develop linearly with the plot, but his Victorian company stays static, and the Eloi and Morlocks regress with the plot. See James Phelan, *Living to Tell about It: A Rhetoric and Ethics of Character Narration* (Ithaca: Cornell University Press, 2005).

33. Patrick Parrinder, "News from Nowhere, The Time Machine, and the Break-Up of Classical Realism," *Science Fiction Studies* 3, no. 3 (November 1976), 272. Adrienne Ghaly remarks that the rise of everyday realist fiction in Wells's time, to which Wells also contributed, allowed writers to convey a detailed phenomenology, especially at the level of food and fashion, of a wide range of experiences of loss, including biodiversity loss. Adrienne Ghaly, "What Does Biodiversity Loss Feel Like? Realism in the Age of Extinction," *New Literary History* 53, no. 1 (2022), 33–57.

34. Fredric Jameson, *Postmodernism, or, The Cultural Logic of Late Capitalism* (Durham: Duke University Press, 1991), 279.

35. Darko Suvin, *Metamorphoses of Science Fiction: On the Poetics and History of a Literary Genre* (New Haven: Yale University Press, 1979), 217.

36. Joanna Russ, *To Write Like a Woman: Essays in Feminism and Science Fiction* (Bloomington: University of Indiana Press, 1995), 7.

37. Wells, *Time Machine*, 87.

38. Jean-Luc Nancy, *The Inoperative Community*, tr. Peter Connor et. al. (Minneapolis: University of Minnesota Press, 1991), 65.

39. Jean-Jacques Rousseau, *The Confessions, Collected Writings of Rousseau*, vol. 5, tr. Christopher Kelly (Hanover: University Press of New England, 1995), 191.

40. Gary K. Wolfe remarks that most extinction narratives do point to some characters or ways of life as surviving in spite of it all: "In fact, most such fictions that we conveniently label 'end of the world' stories are in fact quite the opposite, and dwell on the survival of key representative individuals and in some cases key institutions (such as the family) as well. It might be more accurate to label such fictions 'almost-the-end-of-the-world' fictions, or 'end-of-most-of-the world' fictions, but works that describe a complete annihilation of the planet and all human life are comparatively rare." Gary K. Wolfe, *Evaporating Genres: Essays on Fantastic Literature* (Middletown, CT: Wesleyan University Press, 2011), 105.

41. Wells, *Time Machine*, 93.

42. Jacques Khalip discusses the attention to scenes of the end of the world and lastness as repeatedly invoked and disavowed in the Romantic period in *Last Things: Disastrous Form from Kant to Hujar* (New York: Fordham University Press, 2018).

43. Theodor W. Adorno, *Aesthetic Theory*, tr. Robert Hullot-Kenter (Minneapolis: University of Minnesota Press, 1997), 141.

44. Wai Chee Dimock, *Weak Planet: Literature and Assisted Survival* (Chicago: University of Chicago Press, 2020).

45. Noel Gough, "Playing with Wor(l)ds: Science Fiction as Environmental Literature," in *Literature of Nature: An International Sourcebook*, ed. Patrick D. Murphy (Chicago: Fitzroy Dearborn, 1998), 410.

46. H. G. Wells, *The Outlook for Homo Sapiens* (London: Seeker and Warburg, 1942), 51.

## 4 / Concepts of Extinction in the Holocaust

1. Most historians understand that the decision to enact genocide upon all Jews became fully programmed by 1941 and agreed upon in further detail at the Wannsee Conference on January 20, 1942. For an account of this time period, see Saul Friedländer, *Nazi Germany and the Jews: Volume II, The Years of Extermination, 1939–1945* (New York: Harper Collins, 2007).

2. Daniel Jonah Goldhagen details this eliminationist worldview in *Hitler's Willing Executioners: Ordinary Germans and the Holocaust* (New York: Vintage Books, 1997). Goldhagen cites evidence of dozens of prominent anti-Semitic writers in Germany as early as the time period between 1861 and 1895 who advocated for the physical extermination of the Jews. These early eliminationist arguments established the attitudes that would become the basis for the National Socialist movement. Goldhagen argues: "The elective affinity between the development of the notion of

the unchanging and unchangeable nature of the Jews, conceptualized primarily in explicitly racial terms, and seeing the 'solution' to the 'Jewish Problem' to be their physical annihilation, is unmistakable. The eliminationist mind-set tended towards an exterminationist one" (71). However, some historians have criticized Goldhagen's argument that these writings show evidence of widespread exterminationist intentions across the German public in the decades prior to the Third Reich. For a discussion of these historians' debates, see Yehuda Bauer, *Rethinking the Holocaust* (New Haven: Yale University Press, 2001).

3. Hannah Arendt, *Between Past and Future* (New York: Penguin 2006), 228.

4. Bruno Bettleheim, *Surviving and Other Essays* (New York: Vintage, 1980), 9.

5. For citations from *Mein Kampf*, I use the unabridged edition in a collective translation published by Reynal and Hitchcock: Adolf Hitler, *Mein Kampf* (New York: Reynal and Hitchcock, 1941). Hitler repeatedly linked the vision of potential Aryan extinction to the need to exterminate or extirpate Jews: "The nationalization of our masses will only be successful if, along with all positive fighting for the soul of our people, its international poisoners are extirpated [*ausgerottet*]. All great questions of the times are questions of the moment, and they represent only consequences of certain causes. Only one of them is of causal importance, that is, the question of the racial preservation of the nationality" (469). "He who wants to redeem the German people from the qualities and the vices which are alien to its original nature will have to redeem it first from the alien originators of these expressions. Without the clearest recognition of the race problem and, with it, of the Jewish question, there will be no rise of the German nation" (469). "In this world human culture and civilization are inseparably bound up with the existence of the Aryan. His dying-off [*Aussterben*] or his decline would again lower upon this earth the dark veils of a time without culture" (581).

6. Cited in Hannah Arendt, *The Origins of Totalitarianism* (New York: Harcourt, Brace & Co, 1973), 438. The citation is from Hitler's "Table Talk" collected statements.

7. Cited in Boria Sax, *Animals in the Third Reich* (Providence: Yogh & Thorn Books, 2013), 96.

8. Cited in Benno Müller-Hill, *Murderous Science: Elimination by Scientific Selection of Jews, Gypsies, and Others in Germany, 1933–1945*, tr. G. R. Fraser (Woodbury, NY: Cold Spring Harbor Laboratory Press, 1998), 18.

9. Anson Rabinbach and Sander L. Gilman, *The Third Reich Sourcebook* (Berkeley: University of California Press, 2013), 786. Himmler, on October 4, 1943, during a secret meeting with top SS officials in the city of Poznań (Posen), and on October 6, 1943, in a speech to the party elite, referred explicitly to the "extermination" [*Ausrottung*] of the Jewish people.

10. Rabinbach and Gilman, *Third Reich Sourcebook*, 786.

11. Anne Bäumer-Schleinkofer, *Nazi Biology and Schools*, tr. Neil Beckhaus (New York: Peter Lang, 1995), 30. Along similar lines, Walter Greite wrote in an article for *der Biologe*: "Our National Socialist state leadership and our concept of the people are thoroughly permeated and impregnated with biological trains of thought." Cited in Bäumer-Schleinkofer, 31.

12. Charles Patterson, *Eternal Treblinka: Our Treatment of Animals and the Holocaust* (New York: Lantern Books, 2002); *How Green Were the Nazis? Nature, Environment, and Nation in the Third Reich*, ed. Franz-Josef Bruggemeier, Mark Cioc, and Thomas Zeller (Athens: Ohio University Press, 2005); Frank Uekoetter,

*The Green and the Brown: A History of Conservation in Germany* (Cambridge: Cambridge University Press, 2006); Thomas M. Lekan, *Imagining the Nation in Nature: Landscape Preservation and German Identity, 1885–1945* (Cambridge, MA: Harvard University Press, 2004); Roberto Esposito, *Bios: Biopolitics and Philosophy*, tr. Timothy Campbell (Minneapolis: University of Minnesota Press, 2008); Timothy Snyder, *Black Earth: The Holocaust as History and Warning* (New York: Tim Duggan Books, 2015). The books by Sax and Patterson have received much attention and criticism for their association of animal rights issues today with a history of how Nazism conflated the lives of *untermenschen*, or lowly life, with scorned animal life, seeing both as deserving of eradication. The title of Patterson's book comes from Isaac Bashevis Singer, who writes that, for animals, "all people are Nazis; for the animals it is an eternal Treblinka." Isaac Bashevis Singer, "The Letter Writer," in *The Collected Stories of Isaac Bashevis Singer* (New York: Farrar, Straus and Giroux, 1982), 271. This chapter is not the place to discuss the contemporary research on animal treatment as compared to Jews in the Holocaust. Yet examining animal studies in relationship to the Holocaust can certainly help us understand further what it meant in Nazism to see human and nonhuman animal life, biology, the Volk, and racial purity as intertwined terms definitive of a worldview. For a further discussion of the overlaps of industrial animal operations and Nazism, see David Sztybel, "Can the Treatment of Animals Be Compared to the Holocaust?" *Ethics & the Environment* 11, no. 1 (Spring 2006), 97–132.

13. Tim Cole, "'Nature Was Helping Us': Forests, Trees, and Environmental Histories of the Holocaust," *Environmental History* 19, no. 4 (2014), 665–86; Jacek Małczyńskia, Ewa Domańska, Mikołaj Smykowskic and Agnieszka Kłos, "The Environmental History of the Holocaust," *Journal of Genocide Research* 22, no. 2 (2020), 183–96; Omer Bartov, "What Is the Environmental History of the Holocaust?" *Journal of Genocide Research* 23, no. 3 (2021), 1–10.

14. Ernst Haeckel, *Natürliche Schöpfungsgeschichte: Gemeinverständliche wissenschaftliche vorträge über die entwickelungslehre im allgemeinen und diejenige von Darwin, Goethe und Lamarck im besonderen* (Berlin: Georg Reimer, 1868).

15. The declaration is from Hans Dietrich, NS Reichstagabgeordneter. Cited in the anonymous volume *Der Gelbe Fleck: Die Ausrottung von 500,000 Deutschen Juden* (Paris: Editions du Carrefour, 1936), 237.

16. Friedländer distinguishes racial anti-Semitism from redemptive anti-Semitism: "Whereas ordinary racial anti-Semitism is one element within a wider racist worldview, in redemptive anti-Semitism the struggle against the Jews is the dominant aspect of a worldview in which other racist themes are but secondary appendages. Redemptive anti-Semitism was born from the fear of racial degeneration and the religious belief in redemption." Saul Friedländer, *Nazi Germany and the Jews: Volume I: The Years of Persecution: 1933–1939* (New York: Harper Collins, 1997), 87.

17. Cited in Goldhagen, *Hitler's Willing Executioners*, 143.

18. Ute Deichmann, *Biologists Under Hitler*, tr. Thomas Dunlap (Cambridge, MA: Harvard University Press, 1996), 76.

19. Cited in Müller-Hill, *Murderous Science*, 25.

20. Robert N. Proctor, *Racial Hygiene: Medicine under the Nazis* (Cambridge, MA: Harvard University Press, 1988), 7.

21. For example, in an anonymously written tract on SS marriage policy, one finds the statement: "An average fertility rate has a tremendous impact on the survival of a

race. Where unequal rates of reproduction exist between two or three races living in the same environment, the race with the lower birthrate is marked for extinction after only a few generations." In Rabinbach and Gilman, *Third Reich Sourcebook*, 333.

22. Cited in Bäumer-Schleinkofer, *Nazi Biology and Schools*, 54.

23. Cited in Richard Burkhardt, *Patterns of Behavior: Konrad Lorenz, Niko Tinbergen, and the Founding of Ethology* (Chicago: University of Chicago Press, 2005), 258.

24. Charles Darwin, *On the Origin of Species* (Oxford: Oxford University Press, 1996), 53.

25. Terrence Des Pres, *The Survivor: An Anatomy of Life in the Death Camps* (Oxford: Oxford University Press, 1976), 142.

26. Tzvetan Todorov, *Facing the Extreme: Moral Life in the Concentration Camps*, tr. Arthur Denner and Abigail Pollak (New York: Henry Holt, 1996), 123.

27. Saul Friedländer, "The 'Final Solution': On the Unease in Historical Interpretation," in *The Holocaust: Theoretical Readings*, ed. Neil Levi and Michael Rothberg (New Brunswick: Rutgers University Press, 2003), 71.

28. David Blackbourn, *The Conquest of Nature: Water, Landscape, and the Making of Modern Germany* (New York: Norton, 2007), 46.

29. Wolves have returned to Germany in the last few decades, and the first wolf pack was spotted in 2000.

30. Garry Marvin, *Wolf* (London: Reaktion Books, 2012), 76.

31. Sax, *Animals in the Third Reich*, 66.

32. S. Deinet, C. Ieronymidou, L. McRae, I. J. Burfield, R. P. Foppen, B. Collen, and M. Böhm, *Wildlife Comeback in Europe: The Recovery of Selected Mammal and Bird Species*. Final report to Rewilding Europe by ZSL, BirdLife International and the European Bird Census Council (London: ZSL, 2013).

33. Cited in Friedemann Schmoll, "Indication and Identification: On the History of Bird Protection in Germany 1800–1918," in *Germany's Nature: Cultural Landscapes and Environmental History*, ed. Thomas Lekan and Thomas Zeller (New Brunswick: Rutgers University Press, 2005), 167.

34. Joerg Hartung, "A Short History of Livestock Production," in *Livestock Housing: Modern Management to Ensure Optimal Health and Welfare of Farm Animals*, ed. Andres Aland and Thomas Banhazi (Wageningen: Wageningen Academic Publishers, 2013), 28.

35. Franz Graf Zedwitz, *Die Deutsche Tierwelt*, tr. Jan Lensen (Berlin: Zeitgeschichte Verlag Wilhelm Undermann, 1937), 318.

36. James M. Jasper and Dorothy Nelkin, *The Animal Rights Crusade: The Growth of a Moral Protest* (New York: Free Press 1992), 23.

37. Frank Uekotter, *The Greenest Nation?: A New History of German Environmentalism* (Cambridge, MA: MIT Press, 2014), 54.

38. Thomas Lekan, *Imagining the Nation in Nature*, 173.

39. Martin Heidegger, *Poetry, Language, Thought*, tr. Albert Hofstadter (New York: Harper and Row, 1971), 12.

40. Martin Heidegger, *The Fundamental Concepts of Metaphysics: World, Finitude, Solitude*, tr. William McNeill and Nicholas Walker (Bloomington: University of Indiana Press, 1995), 267.

41. Lutz Heck, *Animals: My Adventure*, tr. E. W. Dickes (London: Methuen, 1954), 156.

42. Fritz Lenz, Erwin Baur, and Eugen Fischer, in the third edition of *Rassenhygiene*, refer to Hitler as "the great German doctor" (cited in Esposito, *Bios*, 112). They add: "We cannot doubt that National Socialism is honestly striving for a healthier race. The question of the quality of our hereditary endowment is a hundred times more important than the dispute over capitalism or socialism" (cited in Müller-Hill, *Murderous Science*, 9). Fischer supported the Nazi eliminationist agenda against Jews for purported biological reasons, declaring in 1939 lecture, "When a people wants, somehow or other, to preserve its own nature, it must reject alien racial elements, and when these have already insinuated themselves, it must suppress them and eliminate them. The Jew is such an alien and, therefore, when he wants to insinuate himself, he must be warded off. This is self-defence. . . . I reject Jewry with every means in my power, and without reserve, in order to preserve the hereditary endowment of my people" (Müller-Hill, 13). Lenz and Fischer continued their work as academic biologists in Germany after the end of the Third Reich. In post-war writings, both claimed that National Socialism had misinterpreted their work. Both Lenz and Fischer declared they were abhorred by the Holocaust in a remarkably frank discussion of Nazism and biology recorded in documents compiled by UNESCO that invited experts across the biological and social sciences to address "The Race Concept." The UNESCO effort led to a 1950 declaration on race and racism, and a further publication with commentary from international academics in 1951. In this 1951 document, Lenz insists it is still scientifically correct to state that humans are a "hybrid community" and should not be called a unified species (he claims the category "human species" is a mistake originating with Linnaeus) and that there remain different "sub-groups" of humans that can interbreed. UNESCO, "The Race Concept" (Paris: UNESCO, 1951), 37. Lenz adds that there are identifiable genetic traits of subgroups and that cultures and generations can still be said to "degenerate," citing the collapse of ancient Greek civilization as an example.

43. Jamie Lorimer and Clemens Driessen, "From 'Nazi Cows' to Cosmopolitan 'Ecological Engineers': Specifying Rewilding through a History of Heck Cattle," *Annals of the American Association of Geographers* 106, no. 3 (2016), 638. See, also, Clemens Driessen and Jamie Lorimer, "Back Breeding the Aurochs: The Heck Brothers, National Socialism, and Imagined Geographies for Non-Human *Lebensraum*," in *Hitler's Geographies: The Spatialities of the Third Reich*, ed. Paolo Giaccaria and Claudio Minca (Chicago: University of Chicago Press, 2016), 138–60.

44. Ronald Goderie, Wouter Helmer, Henri Kerkdijk-Otten, and Staffan Widstrand, *The Aurochs: Born to Be Wild: The Comeback of a European Icon* (Zutphen: Roodbont Publishers, 2013), 117.

45. Goderie et. al., *The Aurochs: Born to Be Wild*, 4.

46. Arendt, *The Origins of Totalitarianism*, 300.

47. Hannah Arendt, *Eichmann in Jerusalem: A Report on the Banality of Evil* (New York: Penguin, 1992), 279.

48. Giorgio Agamben, *Remnants of Auschwitz: The Witness and the Archive*, tr. Daniel Heller-Roazen (New York: Zone Books, 2002), 63.

49. Giorgio Agamben, *Homo Sacer: Sovereign Power and Bare Life*, tr. Daniel Heller-Roazen (Stanford: Stanford University Press, 1998), 137.

50. Agamben, *Remnants of Auschwitz*, 121.

51. Agamben, *Homo Sacer*, 47.

52. Giorgio Agamben, *The Open: Man and Animal*, tr. Kevin Attell (Stanford: Stanford University Press, 2004).

53. Roberto Esposito, *Bíos: Biopolitics and Philosophy*, tr. Timothy Campbell (Minneapolis: University of Minnesota Press, 2008).

54. Robert Antelme, *The Human Race*, tr. Jeffrey Haight and Annie Mahler (Evanston: Marlboro Press/Northwestern University Press, 1998), 218–20.

55. Agamben briefly mentions Antelme in his study of "remnants" and remarks that his work testifies to "a matter of biological belonging in the strict sense," a comment that is rather vague and does not specify how the embedded ecological matters Antelme discusses exceed notions of "strict" or "ultimate" demarcations. Agamben, *Remnants of Auschwitz*, 58.

## 5 / Critical Theory for the Critically Endangered

1. *IUCN Red List Categories and Criteria*, version 3.1, 2nd edition (IUCN, 2012).

2. Michael E Soulé, ed., *Viable Populations for Conservation* (Cambridge: Cambridge University Press, 1987).

3. J. B. MacKinnon, *The Once and Future World: Nature as It Was, as It Is, as It Could Be* (Toronto: Random House Canada, 2013), 38.

4. Richard Mackay, *The Atlas of Endangered Species* (Berkeley: University of California Press, 2008).

5. World Wildlife Fund, *Living Planet Report—2018: Aiming Higher*, ed. Monique Grooten and Rosamunde Almond (Gland: WWF, 2018).

6. In 2002, several marine biologists, divers, and filmmakers set up a media and research campaign at www.shiftingbaselines.org to draw attention to shifting baseline phenomena across oceans. Further discussion of the issue of massively reduced populations now taken as the norm can be found in Caroline Fraser, *Rewilding the World: Dispatches from the Conservation Revolution* (New York: Metropolitan Books, 2009).

7. Michael E. Soulé and L. Scott Mills, "Conservation Genetics and Conservation Biology: A Troubled Marriage," in *Collected Papers of Michael E. Soulé: Early Years in Conservation Biology* (Washington, DC: Island Press, 2014), 189–211.

8. Thom van Dooren, Eben Kirksey, and Ursula Münster, "Multispecies Studies: Cultivating Arts of Attentiveness," *Environmental Humanities* 8, no. 1 (2016), 12.

9. Donna Haraway, *Staying with the Trouble: Making Kin in the Chthulucene* (Durham: Duke University Press, 2016), 35.

10. For a detailed history of the work to repopulate the few remaining bison, see Mark V. Barrow Jr., *Nature's Ghosts: Confronting Extinction from the Age of Jefferson to the Age of Ecology* (Chicago: University of Chicago Press, 2009).

11. Gary Snyder, *The Practice of the Wild* (New York: North Point Press, 1990); Dave Foreman, *Rewilding North America: A Vision for Conservation in the 21st Century* (Washington, DC: Island Press, 2004); Paul S. Martin, *Twilight of the Mammoths: Ice Age Extinction and the Rewilding of American* (Berkeley: University of California Press, 2005); Emma Marris, *Rambunctious Garden: Saving Nature in a Post-Wild World* (New York: Bloomsbury, 2011); George Monbiot, *Feral: Rewilding the Land, Sea, and Human Life* (Chicago: University of Chicago Press, 2014); Marc Bekoff, *Rewilding Our Hearts: Building Pathways of Compassion and Coexistence* (Novato, CA: New World Library, 2014).

12. Gary Snyder, *No Nature: New and Selected Poems* (New York: Pantheon Books, 1992), 356.

13. Ernest Callenbach, *Ecotopia* (New York: Bantam Books, 1975); Ernest Callenbach, *Bring Back the Buffalo! A Sustainable Future for America's Great Plains* (Washington, DC: Island Press, 1996).

14. Thom van Dooren, "Invasive Species in Penguin Worlds: An Ethical Taxonomy of Killing for Conservation," *Conservation & Society* 9, no. 4 (2011), 286–98.

15. Tim Flannery, *The Eternal Frontier: An Ecological History of North America and Its Peoples* (New York: Grove Press, 2001).

16. Ken Zontek, *Buffalo Nation: American Indian Efforts to Restore the Bison* (Lincoln: University of Nebraska Press, 2007), xiii.

17. Winona LaDuke, *All Our Relations: Native Struggles for Land and Life* (Cambridge: South End Press, 1999), 139–66.

18. Nick Estes, *Our History Is the Future: Standing Rock versus the Dakota Access Pipeline and the Long Tradition of Indigenous Resistance* (London: Verso, 2019), 110.

19. Jack Halberstam, *Wild Things: The Disorder of Desire* (Durham: Duke University Press, 2020), x.

20. Foreman, *Rewilding North America*, 3.

21. See "Tiger Species Survival Plan," https://support.mnzoo.org/tigercampaign/tiger-ssp/.

22. Annu Jalais, *Forest of Tigers: People, Politics and Environment in the Sundarbans* (New Delhi: Routledge, 2010), 9.

23. Ronald Sandler, *The Ethics of Species: An Introduction* (Cambridge: Cambridge University Press, 2012), 104.

24. John Vaillant, *The Tiger: A True Story of Vengeance and Survival* (Toronto: Vintage Canada, 2011).

25. Jacqueline Schneider, *Sold into Extinction: The Global Trade in Endangered Species* (Santa Barbara: Praeger, 2012), 80.

26. Vanda Felbab-Brown, *The Extinction Market: Wildlife Trafficking and How to Counter It* (Oxford: Oxford University Press, 2017), 50.

27. Rachel Love Nuwer, *Poached: Inside the Dark World of Wildlife Trafficking* (New York: Da Capo, 2018), 75.

28. Felbab-Brown, *Extinction Market*, 60.

29. Schneider, *Sold into Extinction*.

30. Felbab-Brown, *Extinction Market*, 5.

31. Camilla Calamandrei, dir. *The Tiger Next Door* (Films Transit International, 2009).

32. Peter Laufer details the nexus of ideologies and economies of the exotic pet trade in *Inside the World of Animal Smuggling and Exotic Pets* (Guilford, CT: Lyons Press, 2010).

33. Rosemary-Claire Collard, *Animal Traffic: Lively Capital in the Global Exotic Pet Trade* (Durham: Duke University Press, 2020), 7.

34. Sarah Jaquette Ray, "Environmental Justice and the Ecological Other in Ana Castillo's *So Far from God*," in *Latinx Environmentalisms: Place, Justice, and the Decolonial*, ed. Sarah D. Wald et al. (Philadelphia: Temple University Press, 2019), 156.

35. Neel Ahuja, *Bioinsecurities: Disease Interventions, Empire, and the Government of Species* (Durham: Duke University Press, 2016), x.

36. See George Feldhamer, Joseph Whittaker, Anne-Marie Monty, and Claire Weickert, "Charismatic Mammalian Megafauna: Public Empathy and Marketing Strategy," *Journal of Popular Culture* 36, no. 1 (2002), 160–67; Fiona Sunquist, "Who's Cute, Cuddly and Charismatic?" *International Wildlife* 22 (1992), 4–12.

37. Jamie Lorimer, "Nonhuman Charisma," *Environment and Planning D* 25, no. 5 (2007), 911–32; Monika Krause and Katherine Robinson, "Charismatic Species and Beyond: How Cultural Schemas and Organisational Routines Shape Conservation," *Conservation and Society* 15, no. 3 (2017), 313–21.

38. Marion Endt-Jones, "A Monstrous Transformation: Coral Art in Culture," in *Coral: Something Rich and Strange*, ed. Marion Endt-Jones (Liverpool: Liverpool University Press), 12.

39. Herman Melville, *Typee, Omoo, Mardi* (New York: Library of America, 1982), 488.

40. Quoted in Alastair Sponsel, "From Cook to Cousteau: The Many Lives of Coral Reefs," in *Fluid Frontiers: Exploring Oceans, Islands, and Coastal Environments*, ed. John Gillis and Franziska Torma (Cambridge: White Horse Press, 2015), 153.

41. Rebecca Albright, "Can We Save the Corals?" *Scientific American* 318, no. 1 (2018), 42. A recent study calculated a loss of 50 percent of total coral life since the 1950s. Tyler D. Eddy et al., "Global Decline in Capacity of Coral Reefs to Provide Ecosystem Services," *One Earth* 4, no. 9 (2021), 1278–85.

42. Stefan Helmreich, *Sounding the Limits of Life: Essays in the Anthropology of Biology and Beyond* (Princeton: Princeton University Press, 2015), 60.

43. Haraway, *Staying with the Trouble*, 12.

44. Cecilia Chen, Janine MacLeod, and Astrida Neimanis, "Introduction: Toward a Hydrological Turn?" in *Thinking with Water*, ed. Cecilia Chen, Janine MacLeod, and Astrida Neimanis (Montreal: McGill-Queen's University Press, 2013). In their introduction, the editors of *Thinking with Water* articulate how "thinking with" engages waters as places of locally shared material and cultural histories. By contrast, "Thinking *of* or *about* water in these ways may nonetheless repeat the assumption that water is a resource needing to be managed and organized" (3).

45. J. E. N. Veron, *A Reef in Time: The Great Barrier Reef from Beginning to End* (Cambridge, MA: Harvard University Press, 2009), 221. Iain McCalman reiterates the metaphor: "Corals are indeed the canaries of climate change, and they face death from many more threats than noxious gases in coalmines." Iain McCalman, *The Reef: A Passionate History: The Great Barrier Reef from Captain Cook to Climate Change* (New York: Farrar, Strauss and Giroux, 2013), 275. Consider for a moment the metaphor of the "canary in the coalmine." Where did that canary come from and how did the animal arrive at the mine? There is a biopolitics and a story of global capture and trade of canaries that is implicit in the metaphor and material practice of carrying canaries into coalmines to detect noxious gases.

46. Justin Prystash, "Zoomorphizing the Human: How to Use Darwin's Coral and Barnacles," *Rhizomes*, 24 (2002), n.p., http://rhizomes.net/issue24/prystash/index.html.

47. Coral scientist Charles Birkeland emphasizes the polar opposite qualities of coral that make it all the more difficult to find a way to respond to the dualities of coral's condition. "Although coral reefs are the most productive ecosystems in the sea, the fisheries of coral reefs are among the most vulnerable to overexploitation.

Despite having the power to create the most massive structures in the world made by living creatures (including man), the thin film of living tissue of coral reef is particularly vulnerable to natural disturbances and effects of human activities. . . . This combination of attributes—creative power and fragility, resilience and susceptibility, productivity and vulnerability to overexploitation—makes management of coral-reef systems a particular challenge to science." Charles Birkeland, "Coral Reefs in the Anthropocene," in *Coral Reefs in the Anthropocene*, ed. Charles Birkeland (New York: Springer, 2015), 7.

48. UNESCO, World Heritage List, "The Great Barrier Reef," 2018, https://whc .unesco.org/en/list/154.

49. Veron, *A Reef in Time*, 89.

50. William Shakespeare, *The Tempest* (New Haven: Yale University Press, 2006), 35–36.

51. See Marissa Fessenden, "Sea Coral Makes Excellent Human Bone Grafts," *Smithsonian Magazine* (October 23, 2014), http://www.smithsonianmag.com/smart -news/sea-coral-makes-excellent-human-bone-grafts-180953121/. The consumption of many corals taken orally or injected into the blood stream for medicinal purposes has long been practiced in some Western and non-Western medical traditions.

52. Monique Allewaert, *Ariel's Ecology: Plantations, Personhood, and Colonialism in the American Tropics* (Minneapolis: University of Minnesota Press, 2013).

53. Matthew Griffiths, *The New Poetics of Climate Change: Modernist Aesthetics for a Warming World* (London: Bloomsbury, 2017), 166.

54. Charles Darwin, *Charles Darwin's Notebooks 1836–1844*, ed. Paul Barrett, Peter J. Gautrey, Sandra Herbert, David Kohn, and Sydney Smith (Cambridge: Cambridge University Press, 1987), 177.

55. Prystash, "Zoomorphizing the Human," n.p.

56. Charles Darwin, *Journal of Researches into the Natural History and Geology of the Countries Visited during the Voyage of H.M.S. Beagle Round the World, under the Command of Capt. Fitz Roy, R.N.*, 2nd edition (London: John Murray, 1845), 460.

57. Jules Verne, *Twenty Thousand Leagues under the Sea*, tr. William Butcher, (Oxford: Oxford University Press, 1998), 173.

58. Toni M. Gregg, Lucas Mead, John H. R. Burns, and Misaki Takabayashi, "Puka Mai He Ko'a: The Significance of Corals in Hawai'ian Culture," in *Ethnobiology of Corals and Coral Reefs*, ed. Nemer Narchi and Lisa Price (New York: Springer, 2015), 107.

59. Epeli Hau'ofa, *We Are the Ocean: Selected Works* (Honolulu: University of Hawai'i Press, 2008), 37.

60. Jeff Orlowski, dir. *Chasing Coral* (Netflix, 2017).

61. Rowan Jacobsen, "Obituary: Great Barrier Reef (25 Million BC- . . .)," *Outside* (October 11, 2016), https://www.outsideonline.com/2112086/obituary-great-barrier -reef-25-million-bc-201.

62. Cited in McCalman, *The Reef*, 244.

63. On the DVD edition, after each episode of *Planet Earth*, however, the producers added a further featurette that revealed how some scenes were filmed. For a longer history of wildlife documentary film that focuses on the popularization of this genre in the United States, see Gregg Mittman, *Reel Nature: America's Romance with Wildlife on Film* (Cambridge, MA: Harvard University Press, 1999).

64. Stacy Alaimo, "The Anthropocene at Sea: Temporality, Paradox, Compression," in *The Routledge Companion to the Environmental Humanities*, ed. Ursula Heise, Jon Christensen, and Michelle Niemann (New York: Routledge, 2017), 158.

65. Bärbel G. Bischof, "Geographies of Coral Reef Conservation: Global Trends and Environmental Constructions," in *Water Worlds: Human Geographies of the Ocean*, ed. Jon Anderson and Kimberley Peters (London: Routledge, 2016), 51–72.

66. Irus Braverman, *Coral Whisperers: Scientists on the Brink* (Berkeley: University of California Press, 2018), 14.

## 6 / What Is De-Extinction?

1. Stewart Brand, "The Dawn of De-Extinction: Are You Ready?" TEDx (2013), https://www.ted.com/talks/stewart_brand_the_dawn_of_de_extinction_are_you _ready/transcript?language=en#t-405473.

2. J. M. Barrie, *Peter Pan* (Mineola, NY: Dover, 2000), 52.

3. Charis Thompson, *Making Parents: The Ontological Choreography of Reproductive Technologies* (Cambridge, MA: MIT Press, 2005).

4. For more on Benirschke's work, see Kurt Benirschke, "The Frozen Zoo Concept," *Zoo Biology* 3, no. 4 (1984), 325–28. Benirschke notes in this essay, "You must collect things for reasons we don't understand" (326). See, also, Oliver A. Ryder and Kurt Benirschke, "The Potential Use of 'Cloning' in the Conservation Effort," *Zoo Biology* 18 (1997), 295–300.

5. Donna Haraway, *When Species Meet* (Minneapolis: University of Minnesota Press, 2008), 252.

6. Oliver Ryder, "Opportunities and Challenges for Conserving Small Populations: An Emerging Role for Zoos in Genetic Rescue," in *The Ark and Beyond: The Evolution of Zoo and Aquarium Conservation*, ed. Ben A. Minteer, Jane Maienschein, and James P. Collins (Chicago: University of Chicago Press, 2018), 261.

7. Joanna Radin and Emma Kowal, "Freezing Politics," in *Cryopolitics: Frozen Life in a Melting World*, ed. Joanna Radin and Emma Kowal (Cambridge, MA: MIT Press, 2017), 8.

8. Nikolas Rose, *The Politics of Life Itself: Biomedicine, Power, and Subjectivity in the Twenty-First Century* (Princeton: Princeton University Press, 2007), 7.

9. Robert Mitchell and Catherine Waldby, "National Biobanks: Clinical Labor, Risk Production, and the Creation of Biovalue," *Science, Technology, & Human Values* 35, no. 3 (May 2010), 330–55.

10. Michelle Murphy, *Seizing the Means of Reproduction: Entanglements of Feminism, Health, and Technoscience* (Durham: Duke University Press, 2012), 1.

11. Douglas Ian Campbell and Patrick Michael Whittle, *Resurrecting Extinct Species: Ethics and Authenticity* (Cham, Switzerland: Palgrave Macmillan, 2017), 5.

12. Matthew Chrulew, "Freezing the Arc: The Cryopolitics of Endangered Species Preservation," in *Cryopolitics*, 292.

13. *IUCN SSC Guiding Principles on Creating Proxies of Extinct Species for Conservation Benefit* (Gland, Switzerland: IUCN Species Survival Commission, 2016), 1.

14. Beth Shapiro, *How to Clone a Mammoth* (Princeton: Princeton University Press, 2015), 130.

15. Ben Novak, "De-Extinction," *Genes* 9, no. 11 (2018), 5.

16. Curt Meine, "De-Extinction and the Community of Being," in *Recreating the Wild: De-Extinction, Technology, and the Ethics of Conservation*, Hastings Center Report 47.4 (2017), S13.

17. Ronald Sandler, "The Ethics of Reviving Long Extinct Species," *Conservation Biology* 28, no. 2 (2013), 356.

18. George Church and Ed Regis, *Regenesis: How Synthetic Biology Will Reinvent Nature and Ourselves* (New York: Basic Books, 2012), 53.

19. Rose, *The Politics of Life Itself*, 16.

20. For further discussion, see Joshua Schuster and Derek Woods, *Calamity Theory: Three Critiques of Existential Risk* (Minneapolis: University of Minnesota Press, 2021).

21. Deborah Bird Rose, *Wild Dog Dreaming: Love and Extinction* (Charlottesville: University of Virginia Press, 2011), 44.

22. Deborah Bird Rose, *Shimmer: Flying Fox Exuberance in Worlds of Peril* (Edinburgh: Edinburgh University Press, 2022), 125.

23. Michel Foucault, *The History of Sexuality, Vol. 1: An Introduction*, tr. Robert Hurley (New York: Vintage, 1990), 143.

24. Michel Foucault, *Security, Territory, Population: Lectures at the Collège De France, 1977–78*, tr. Graham Burchell (New York: Palgrave, 2009), 1.

25. Michel Foucault, *"Society Must Be Defended": Lectures at the Collège De France, 1975–76*, tr. David Macey (New York: Picador, 2003), 246–47.

26. Foucault, *The History of Sexuality*, vol. 1, 138.

27. See *Foucault and Animals*, ed. Matthew Chrulew and Dinesh Wadiwel (Leiden: Brill, 2017); Achille Mbembe, *Necropolitics* (Durham: Duke University Press, 2019); Cary Wolfe, *Before the Law: Humans and Animals in a Biopolitical Frame* (Chicago: University of Chicago Press, 2013); Jeffrey T. Nealon, *Plant Theory: Biopower and Vegetable Life* (Palo Alto: Stanford University Press, 2015).

28. Claire Jean Kim, *Dangerous Crossings: Race, Species, and Nature in a Multicultural Age* (Cambridge: Cambridge University Press, 2015), 18.

29. Elizabeth A. Povinelli, *Geontologies: A Requiem to Late Liberalism* (Durham: Duke University Press, 2016), 5.

30. Mbembe, *Necropolitics*; Alexander Weheliye, *Habeas Viscus: Racializing Assemblages, Biopolitics, and Black Feminist Theories of the Human* (Durham: Duke University Press, 2014); Tiffany Lethabo King, *The Black Shoals: Offshore Formations of Black and Native Studies* (Durham: Duke University Press, 2019); Zakiyyah Iman Jackson, *Becoming Human: Matter and Meaning in an Anti-Black World* (New York: New York University Press, 2020); Glen Sean Coulthard, *Red Skin, White Masks: Rejecting the Colonial Politics of Recognition* (Minneapolis: University of Minnesota Press, 2014); Jodi A. Bird, *The Transit of Empire: Indigenous Critiques of Colonialism* (Minneapolis: University of Minnesota Press, 2011).

31. *IUCN SSC Guiding Principles*, 4.

32. Matthew A. Taylor, "Living after Life," *American Literary History* 33, no. 3 (2021), 553.

33. Hans Jonas, *The Phenomenon of Life: Toward a Philosophical Biology* (Evanston: Northwestern University Press, 2001), xxiii.

34. Hans Jonas, *The Imperative of Responsibility: In Search of an Ethics for the Technological Age* (Chicago: University of Chicago Press, 1984), 125.

35. Amy Lynn Fletcher, *De-Extinction and the Genomics Revolution: Life on Demand* (Cham, Switzerland: Palgrave, 2020), 3.

36. S. D. Chrostowska, *Utopia in the Age of Survival: Between Myth and Politics* (Stanford: Stanford University Press, 2021), 71.

37. Meine, "De-Extinction and the Community of Being," S9–10.

38. All citations from the trilogy are from Octavia E. Butler, *Lilith's Brood* (New York: Grand Central Publishing, 2000).

39. Elizabeth Hennessey, "Cryogenic Freezer Box," in *Future Remains: A Cabinet of Curiosities for the Anthropocene*, ed. Gregg Mitman, Marco Armiero, and Robert S. Emmett (Chicago: University of Chicago Press, 2017), 114.

40. Gerry Canavan argues that the Oankali are so coercive and dominant of human bodies and so intent on capturing their genomes that the Oankali can be charged with genocidal acts. Gerry Canavan, *Octavia E. Butler* (Urbana: University of Illinois Press, 2016), 86.

41. Michael P. Marchetti and Peter B. Moyle, *Protecting Life on Earth: An Introduction to the Science of Conservation* (Berkeley: University of California Press, 2010), 37.

42. Jessie Stickgold-Sarah, "'Your Children Will Know Us, You Never Will': The Pessimistic Utopia of Octavia Butler's Xenogenesis Trilogy," *Extrapolations: A Journal of Science Fiction and Fantasy* 51, no. 3 (2010), 425.

43. Hoda M. Zaki, "Utopia, Dystopia, and Ideology in the Science Fiction of Octavia Butler," *Science Fiction Studies* 17, no. 2 (1990), 239–51.

44. Sherryl Vint, *Bodies of Tomorrow: Technology, Subjectivity, Science Fiction* (Toronto: University of Toronto Press, 2007), 67. As Vint notes, "The Oankali may be genetic essentialists, but Butler's readers are encouraged not to be" (67).

45. Nancy Jesser, "Blood, Genes and Gender in Octavia Butler's *Kindred* and *Dawn*," *Extrapolation: A Journal of Science Fiction and Fantasy* 43, no. 1 (2002), 39.

46. Octavia E. Butler, *Conversations with Octavia Butler*, ed. Consuela Francis (Jackson: University Press of Mississippi, 2010), 105.

47. Evelyn Fox Keller, *Refiguring Life: Metaphors of Twentieth-Century Biology* (New York: Columbia University Press, 1995), xv.

48. Sherryl Vint, *Animal Alterity: Science Fiction and the Question of the Animal* (Liverpool: Liverpool University Press, 2010), 158.

49. Gilles Deleuze, *Pure Immanence: Essays on A Life* (New York: Zone Books, 2005), 28.

50. Deleuze, *Pure Immanence*, 27.

51. Cary Wolfe, *Before the Law: Humans and Other Animals in a Biopolitical Frame* (Chicago: University of Chicago Press, 2012), 52.

## Conclusion

1. Fred Adams and Greg Laughlin, *The Five Ages of the Universe: Inside the Physics of Eternity* (New York: Touchstone, 1999).

2. Carl Sagan, *Pale Blue Dot: A Vision of the Human Future in Space* (New York: Ballentine, 1994), 173.

3. Stephen Hawking, https://bigthink.com/u/stephenhawking.

4. Ross Andersen, "Exodus," *Aeon Magazine* (September 30, 2014), https://aeon.co/essays/elon-musk-puts-his-case-for-a-multi-planet-civilisation.

5. Nick Bostrom, "Astronomical Waste: The Opportunity Cost of Delayed Technological Development," *Utilitas* 15, no. 3 (2003), 308–14.

6. Kim Stanley Robinson, *Aurora* (New York: Orbit, 2015).

7. See www.iucnredlist.org/species/17975/123809220.

8. Trevor Paglen, *The Last Pictures* (Berkeley: University of California Press, 2012).

9. Carl Sagan, *Murmurs of Earth: The Voyager Interstellar Record* (New York: Random House, 1978), 236.

10. Walter Benjamin, "On the Concept of History," in *Selected Writings*, vol. 4, tr. Harry Zohn, ed. Michael W. Jennings (Cambridge, MA: Harvard University Press, 2003), 393.

11. Paglen, *The Last Pictures*, xiii.

12. Hannah Arendt, *The Human Condition* (Chicago: University of Chicago Press, 1958), 168.

13. Jacques Derrida, *Archive Fever: A Freudian Impression*, tr. Eric Prenowitz (Chicago: University of Chicago Press, 1995), 19.

14. Derrida, *Archive Fever*, 33–34.

15. Paglen, *Last Pictures*, 12.

16. Hans Jonas, *The Imperative of Responsibility: In Search of an Ethics for the Technological Age* (Chicago: University of Chicago Press, 1984), 21.

17. Eileen Crist, *Abundant Earth: Toward an Ecological Civilization* (Chicago: University of Chicago Press, 2019), 63.

# Index

Joshua Schuster is an associate professor of English and core faculty member of the Centre for the Study of Theory and Criticism at Western University. He is the author of *The Ecology of Modernism: American Environments and Avant-Garde Poetics* and co-author of *Calamity Theory: Three Critiques of Existential Risk.*

www.ingramcontent.com/pod-product-compliance
Lightning Source LLC
Jackson TN
JSHW081313130125
77033JS00007B/282